ELEMENTARY
WAVE MECHANICS

ELEMENTARY
WAVE MECHANICS

WITH APPLICATIONS TO
QUANTUM CHEMISTRY

BY

W. HEITLER
PROFESSOR OF THEORETICAL PHYSICS IN THE
UNIVERSITY OF ZÜRICH

SECOND EDITION

OXFORD
AT THE CLARENDON PRESS

Oxford University Press, Amen House, London E.C.4

GLASGOW NEW YORK TORONTO MELBOURNE WELLINGTON
BOMBAY CALCUTTA MADRAS KARACHI KUALA LUMPUR
CAPE TOWN IBADAN NAIROBI ACCRA

FIRST EDITION 1945

REPRINTED 1946 (TWICE), ALSO LITHOGRAPHICALLY
IN GREAT BRITAIN AT THE UNIVERSITY PRESS, OXFORD
FROM SHEETS OF THE THIRD IMPRESSION 1947, 1948, 1950
SECOND EDITION 1956
REPRINTED LITHOGRAPHICALLY 1958

PREFACE TO THE SECOND EDITION

As this little book was primarily designed for the use of chemists and other non-mathematical readers I have added a section on diatomic molecules and above all much extended the chapters on the chemical bond. These are in the same elementary style as the rest of the book. I hope that these additions (which seem compatible with the idea of a small pocket book) will afford a good illustration of the methods of wave mechanics as well as its usefulness for chemical problems. To make the general theory more complete a section on the time-dependent wave equation has also been added.

W. H.

Zürich
January 1956

CONTENTS

I. EXPERIMENTAL BASIS OF QUANTUM MECHANICS
1. Quantum states and electron diffraction   1
2. Relations between wave and particle properties   3
3. Reconciliation of wave and particle views   8

II. DERIVATION OF THE WAVE EQUATION
1. The free electron   17
2. Discrete quantum states   21
3. The Schrödinger wave equation   24
4. The time-dependent wave equation   27

III. THE HYDROGEN ATOM
1. The ground state   32
2. Excited states   37
3. p states   39
4. Normalization and linear combination of wave functions   43

IV. ANGULAR MOMENTUM. ZEEMAN-EFFECT. SPIN
1. Sharp and unsharp quantities   46
2. Angular momentum of s and p states   50
3. Zeeman-effect   54
4. d states, directional quantization   56
5. The electron spin   59
6. Two electrons with spin   64

V. PROBLEM OF TWO ELECTRONS
1. The wave equation for two electrons   68
2. Solution of the wave equation for two electrons   70
3. Exchange degeneracy   74
4. The Pauli exclusion principle   76
5. The spin wave function   79
6. General formulation of the Pauli principle   82

VI. PERTURBATION THEORY
1. General theory — 86
2. He atom, exchange energy — 88
3. The orthogonality theorem — 93

VII. THE PERIODIC SYSTEM OF ELEMENTS
1. The electron configuration — 97
2. The atomic states — 100

VIII. DIATOMIC MOLECULES
1. The electronic states — 107
2. The rotation of molecules — 112
3. The vibration of molecules — 115
4. Ortho- and para-molecules — 117

IX. THEORY OF HOMOPOLAR CHEMICAL BOND
1. The hydrogen molecule — 123
2. The saturation properties of the chemical bond — 136

X. VALENCY
1. Spin valency — 139
2. Crossing of atomic interaction curves. Valency of carbon — 143
3. Interaction in diatomic molecules — 150

XI. POLYATOMIC MOLECULES
1. Interaction of several atoms with one valency electron — 155
2. Activation energy, non-localized bonds — 164
3. Directed valencies — 170
4. Interaction of atoms with several electrons — 179
5. Binding energies of hydrocarbons — 184

INDEX — 191

I

EXPERIMENTAL BASIS OF QUANTUM MECHANICS

1. Quantum states and electron diffraction

THERE are two main groups of experimental phenomena which are inconsistent with classical physics, namely:

(i) when the internal energy of an atom changes, owing to emission or absorption of light, it does not do so evenly or continuously but in 'quantum' steps;

(ii) the fact that a beam of electrons exhibits interference phenomena similar to those of a light wave.

(i) *Quantum states*

A great wealth of experimental material, chiefly derived from spectroscopy, shows that an atom cannot exist in states of continuously varying energy but only in different discrete states of energy, symbolized in Fig. 1, and referred to as 'discrete energy levels'. These levels and the spacing between them are different for the various chemical elements, but are identical for all atoms of the same element. The change from one level E_2 to another E_1 is associated with the emission (if $E_2 > E_1$) or absorption (if $E_2 < E_1$) of light, whose frequency ν is determined by the relation

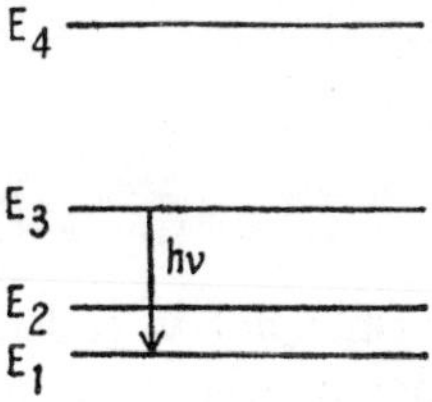

$$E_2 - E_1 = h\nu. \qquad (1)$$

FIG. 1. Discrete energy levels.

h is a universal constant known as Planck's quantum of action whose value is $6{\cdot}6\times 10^{-27}$ erg sec. The frequency for the same energy change, e.g. E_2 to E_1, is always the same. Furthermore, a beam of light with frequency ν, such as may be emitted by a large number of like atoms all carrying out quantum jumps from E_2 to E_1 will not have a continuously varying energy but will consist of a number of 'quanta' each having an energy $h\nu$. The energy of the beam is thus either $1h\nu$ or $2h\nu$ or $3h\nu$... or $nh\nu$, etc., or, in other words, consists of 1, 2, 3,..., n,... light quanta.

(ii) *Interference of waves*

It is well known that light waves when reflected or refracted from a regularly spaced grating interfere and

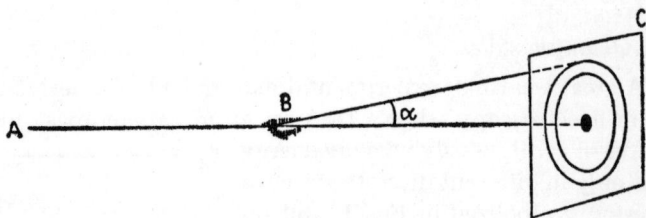

FIG. 2. Electron diffraction.

form what is called a 'diffraction' pattern. This phenomenon was used, for instance, to prove the wave nature of X-rays. For this purpose the regular spacing is provided by the atoms of a crystal, as an ordinary grating would be too coarse. The crystal may be used in the form of a powder. The waves from their source A (Fig. 2) pass through the crystalline powder B and arrive on the screen C. The diffraction pattern observed on C consists of a series of concentric rings (Debye–Scherrer rings) from whose spacing we can determine the wave-length of the light using the formula

$$n\lambda = 2d \sin\frac{\alpha}{2},$$

where d is the distance apart of the rows of atoms in the crystal and α the angle of *diffraction*. $n = 1, 2, 3,...$ for the various rings.

We can now carry out a similar experiment with a beam of *electrons*, all having the same velocity, v, and it has been found that the *result is similar*. To produce the beam of electrons a cathode-ray tube may be used. The electrons pass through an electrical potential difference V which accelerates the electrons to a velocity, v, determined by the relation $eV = \frac{1}{2}mv^2$. The electrons then pass through the crystal powder B, etc., as above.

The diffraction pattern so found is very similar to that produced by X-rays. This experiment shows that *waves are associated with a beam of electrons*, formerly considered as consisting of particles only. It is the purpose of wave mechanics to show how the properties of an electron regarded as a wave can be reconciled with its properties as a particle. At the same time it will be seen that the facts of the discrete energy levels follow from the properly developed theory of wave mechanics.

2. Relations between wave and particle properties

To develop the theory of wave mechanics we start from the experiment described above, showing that a wave is associated with a beam of electrons. The first question which naturally arises is: What is the wave-length λ found to be? It can be measured from the spacing of the diffraction rings. The measurements show that λ depends on the velocity, v, of the electrons. This velocity can be varied by varying the potential used to accelerate the electrons.

The wave-length is then found to be inversely proportional to the velocity; $\lambda \propto 1/v$, the slower the electrons the longer is the wave-length. The proportionality constant can also be measured in this way. It turns out to be equal to Planck's constant divided by the mass of the electron m:

$$\lambda = \frac{h}{mv} \quad \text{or} \quad \lambda = \frac{h}{p}, \tag{2}$$

where mv is called the momentum p. The relation (2) is due to de Broglie (1924) and marks the starting-point of wave mechanics.

The velocity, v (henceforward called the 'particle velocity'), is a concept relevant to the electron pictured as a particle, whilst λ is a concept relevant to a wave. The two concepts are connected by Planck's constant h. This constant was first discovered as giving the connexion between the energy E and the frequency ν of a light quantum,

$$E = h\nu.$$

Whilst the frequency is a concept relevant to a wave, a light quantum rather resembles the concept of a particle. Now we have seen that the same constant h also connects the wave picture with the particle picture for electrons.

A wave is not determined by its wave-length alone but by the three quantities, wave-length λ, frequency ν, and wave velocity v_ϕ which are connected by the relation

$$\lambda\nu = v_\phi. \tag{3}$$

In the case of a light wave $v_\phi = c$ (velocity of light).

We have already found a connexion between λ and the particle velocity v, but in order to determine the properties of the electron waves completely one further relation is

EXPERIMENTAL BASIS OF QUANTUM MECHANICS 5

needed for either ν or v_ϕ. Unfortunately both ν and v_ϕ cannot be determined experimentally. Instead, we shall establish a further relation on theoretical grounds. It is plausible to assume that there must be a connexion between the particle velocity v and the velocity of propagation v_ϕ of the wave. It would be wrong, however, to

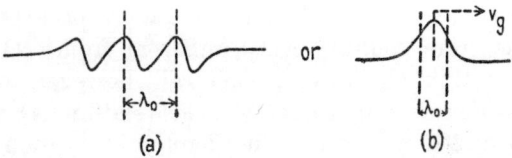

Fig. 3. Group velocity.

assume straight away that $v = v_\phi$. The reason is the following: In our case λ was seen to depend strongly on v and therefore—presumably—also on v_ϕ. Our waves thus show a strong *dispersion*, similar to, but much stronger than, the dispersion of a light wave in a refracting medium, glass or water, say. In this case one distinguishes between *two* velocities of a wave. In addition to the *phase velocity* v_ϕ defined above, one can define a so-called *group velocity*

$$v_g = \frac{d\nu}{d(1/\lambda)}. \qquad (4)$$

Only in the case where v_ϕ is independent of the wavelength λ, is v_g, according to (3), identical with v_ϕ. Whilst the phase velocity is the velocity of propagation of a monochromatic, infinitely long wave, a closer examination shows that the group velocity describes the velocity of travelling of a short wave pulse (Fig. 3). Such a pulse, often also called wave packet, does not have a well-defined wave-length but is composed of waves with many different wave-lengths, centred about the basic wave-length λ_0, just as a pulse of sound (noise) is composed of many different tones. Each of these monochromatic partial waves of which the wave packet is composed has a different phase velocity. It can now be shown that this has the following result: Supposing the wave packet has the shape of Fig. 3 (a) or (b) at the time $t = 0$, then for a short time afterwards the shape is retained and the whole wave packet travels with the group velocity $v_g(\lambda_0)$, different from $v_\phi(\lambda_0)$. (In course of time the wave packet will, however, alter its shape.) For a simple example the relation (4) will be derived and demonstrated in Chapter II, section 4. Of course, in the case of light pulses where dispersion is small, v_ϕ and v_g are *almost* equal. (*In vacuo* they are even exactly equal.)

Now in the case of electron waves we have seen that the particle velocity v depends on the wave-length very strongly, $v \propto 1/\lambda$, showing a very big dispersion. Accordingly, v_ϕ and v_g are very different. If we wish to equate the particle velocity v to the velocity of propagation of the waves we have to decide whether to assume $v = v_\phi$ or $v = v_g$. Since a particle resembles a small wave packet rather than an infinitely long monochromatic wave, it would seem more likely that

$$v = v_g. \tag{5}$$

This would be so at any rate, if we picture a 'particle' as a wave packet of small extension, which, as will be seen later, is correct. Our choice $v = v_g$ will be seen to be confirmed by all experimental facts.

We thus assume that $v_g = v$ or

$$v = \frac{d\nu}{d(1/\lambda)} = \frac{d\nu}{d(mv/h)} = \frac{h}{m}\frac{d\nu}{dv}. \tag{6}$$

We can use this to determine ν, since on integration

$$\nu = \frac{m}{h}\int v\,dv = \frac{1}{h}\frac{m}{2}v^2,$$

or $\quad\quad\quad\quad h\nu = \tfrac{1}{2}mv^2 = E,\quad\quad\quad\quad (7)$

which is the energy E of a particle of mass m and velocity v. Remembering that $h\nu = E$ was also the energy of a light quantum with frequency ν, we now see that the same relation prevails between the energy of the electron considered as a particle and the frequency of the electron waves. The relation $E = h\nu$ is thus universal, and this consistency may be regarded as a confirmation of our choice $v = v_g$.†

We now have a complete connexion between the particle properties of a beam of electrons (velocity, energy) and its wave properties (wave-length, frequency):

Particle view	Wave concept
Velocity $= v$ (momentum $p = mv$)	$\lambda = \dfrac{h}{mv} = \dfrac{h}{p}$
Energy $= \tfrac{1}{2}mv^2$	$\nu = \dfrac{1}{h}\dfrac{1}{2}mv^2 = \dfrac{E}{h}$
	$v_g = v.$

We can also express the phase velocity by the particle velocity $v_\phi = \nu\lambda = \tfrac{1}{2}v$, but this relation is not of great importance. In fact the phase velocity hardly has a physical meaning. Like the frequency with which it is connected by $\nu\lambda = v_\phi$, it is not a measurable quantity (compare footnote). This is also illustrated by the following fact which

† In (7) an additive constant of integration has been omitted. This would mean adding a constant to the energy; this has no physical meaning. Accordingly, the frequency of the electron wave is only determined to within an additive constant (in contrast to the wavelength λ) and is not a measurable quantity.

we mention without proof. If we were to treat our electron waves according to the principles of special relativity, the phase velocity would turn out to be not $v/2$ but c^2/v. It would thus be larger than the velocity of light c, but no velocity larger than c can be measured. The phase velocity has therefore a purely mathematical significance. This, of course, is in contrast to light waves where both the frequency ν and the velocity c have a well-defined meaning.

3. Reconciliation of wave and particle views

We have seen that a beam of electrons must be described 'partly' as consisting of a number of individual particles and partly as a wave. As yet the two concepts stand quite apart and seem irreconcilable. Speculations as to what the 'medium' of the wave is, have proved fruitless.

We must now seek a better and deeper understanding of the mutual relationship between the wave and particle concepts. For this purpose we consider the *intensity* of the electron beam. It is well known that the intensity of a wave at each point of space is proportional to the square of the *wave amplitude*, ψ say. We may regard ψ^2 as a measure of the intensity of the electron wave. ψ will in general be a function of space and time. On the other hand, the local intensity of the beam is equal to the number of particles per cm.[3] For an intense beam, as used above, we therefore equate ψ^2 to the number of electrons per cm.[3] in the beam. The apparent contradiction between the two pictures becomes now very striking if we consider a *single* electron. What can the meaning of ψ^2 be then? How can we conceive a particle, such as exists in an atom, to be associated with or described as a wave? Is the electron a particle and a wave at the same time?

The answer to these questions can be given by an

experiment. We carry out the same diffraction experiment (Fig. 2) with *single* electrons, letting the individual electrons pass one after another through the crystal powder B and record their arrival on a scintillation screen C. If each individual electron were in fact a complete wave we should expect that each time an electron arrives on the screen the whole diffraction pattern with all the rings would appear simultaneously—which is hardly conceivable, as the rings have macroscopic extension! This is not indeed what happens. Instead, we observe that the individual electrons arrive on the screen at individual points scattered over the whole screen. However, when a number of electrons have arrived, we find that the distribution of the points is not uniform, they lie preferably at such places where in the former experiment—using an intense beam of electrons—the maxima of the diffraction pattern would lie. No electrons, or very few, are found to arrive at the minima of the diffraction pattern (Fig. 4: the left-hand diagram shows ψ^2 on the plane of the screen, the right-hand graph shows ψ^2 for a radial cross-section through the screen.) In course of time—when a very large number of electrons have arrived—the rings are built up completely, as in the former experiment with an intense beam.

It is evidently not possible to predict where any one individual electron would appear on the screen (although it is possible to say where an electron would *not* appear, namely, between the rings). This leads us to interpret the intensity ψ^2 (on the screen, plotted in Fig. 4) as the *probability of a single electron hitting the screen at a certain point*. Of course, if a large number of electrons is used, each will be at a certain point with a probability ψ^2, and we are led back to the former interpretation of ψ^2 as the intensity or average density of particles. More generally, ψ^2 as a function of space, will be the *probability for the electron to be present at this particular point of space* (at the time at which the wave function is considered). The probability interpretation of ψ^2 for single electrons is due to Born (1926).

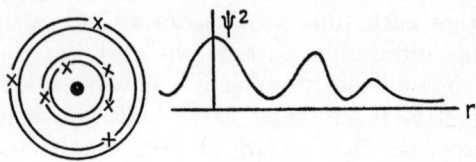

FIG. 4. Probability distribution.

This has now a very important and far-reaching consequence. In contrast to classical mechanics we *cannot predict exactly which way the electron will go*. We can only tell that the probability for finding the electron at a certain place, at a certain time, is, say, 50 or 3 per cent., but we cannot say where the electron will be found precisely. This probability is given by the square of the wave amplitude ψ^2, which, naturally, is a function of time and space.

It follows from the above that *no definite value* can be attached to the position of the electron at any time. Its position may be anywhere within the extension of the wave by which it is described. We say in such a case that the position of the electron is 'uncertain' or 'not sharp' to distinguish it from something which has a 'sharp' value, such as, for instance, its velocity in the above diffraction experiment. The fact that not always can *sharp values be attached to every physical quantity* (in our case to the position) *is a fundamental feature of quantum mechanics or wave mechanics*. It is, as we have seen, due to the *double nature of the electron, as a particle and as a wave*, as it is exhibited by numerous experiments, of which the above diffraction experiment is only one example. The necessity of reconciling these two 'natures' leads directly to the probability interpretation of ψ^2 and to the fundamental principle of the uncertainty of physical quantities.

This principle of uncertainty will also hold for other physical quantities such as, for instance, the angular momentum, etc. We cannot always attach sharp values to a given physical quantity; if a quantity does not have a sharp value, we can only give the probability for its having certain values. On the other hand, there are of course physical quantities, which may have a *sharp* value in certain cases. An example is the wave-length in the diffraction experiment discussed above. Since $\lambda = h/p$, it follows that p (or v) also has a sharp value (it is measured by the applied potential). In addition, the energy for such a beam of electrons is a function of the velocity only, so that it also has a sharp value. We shall see that it depends on the *external conditions*, especially on the *experiment performed*, which physical quantities have sharp values and which have not. In the diffraction experiment the situation is:

Sharp   *Unsharp*

$$p, E = \frac{m}{2} v^2 = \frac{p^2}{2m}. \qquad x, y, z.$$

Hence there are physical quantities with sharp values and others for which we can only give probability values. Another example of a case where the energy is sharp and the position is unsharp is provided by an ordinary atom which shows discrete energy values E_1, E_2, E_3, etc. We are able to state the energy value of the atom in a particular instance, yet the position of the electron is not sharp, although a 'probability distribution' for the various positions of the electron can be given. It will be derived in Chapter III.

A small wave packet also has sharp and unsharp quantities (Fig. 3 (a) or (b)). Let us contrast a plane monochromatic wave, which extends indefinitely and has a fixed velocity, with a very small wave packet. Such a small wave packet has a small but finite extension Δx, say, but as we shall see, the velocity is not defined.

A wave packet can, in fact, be regarded as being built up by a superposition of monochromatic waves in much the same way as a short pulse of sound is built up by a spectrum of monochromatic tones. Therefore a wave packet consists of waves of several wave-lengths, λ varying, say, between $\lambda_0+\Delta\lambda$ and $\lambda_0-\Delta\lambda$. The momentum varies accordingly between $p_0+\Delta p$ and $p_0-\Delta p$; $p_0 = h/\lambda_0$. Hence the *velocity* or *momentum cannot be sharp*, in contrast to a monochromatic wave which has infinite extension but sharp wave-length and momentum. For a wave packet the energy $E = p^2/2m$ is also unsharp. We call Δp the uncertainty of the momentum. The smaller the spatial extension, Δx, of the wave packet, the wider is the range of wave-lengths required to build up the wave packet, and the larger, therefore, is the uncertainty of the momentum Δp.† Now Δx is a measure for the uncertainty of the position, since ψ^2 is different from zero only within the extension of the wave packet. Δx may be very small, in which case the position is nearly sharp. A monochromatic wave has infinite extension $\Delta x = \infty$, but sharp momentum $\Delta p = 0$. Thus we find the following theorem.

The more accurately the position of a particle is defined the less accurate is its momentum or velocity (small wave packet), and the more accurately the momentum of the

† The group velocity v_g (p. 5) is the average of the various velocities within $\Delta v = \Delta p/m$.

particle is defined the less accurate is the position (monochromatic wave). This is *Heisenberg's uncertainty relation* (1927).

These considerations can be carried out quantitatively, calculating the range of wave-lengths of Δp required to build up a packet of size Δx. It is found that

$$\Delta x \Delta v = \frac{h}{2\pi m} \quad \text{or} \quad \Delta x \Delta p = \frac{h}{2\pi}, \tag{8}$$

where h is again Planck's constant. The formula demonstrates the uncertainty principle directly; the product of the uncertainties is a constant proportional to h. A monochromatic wave is a limiting case $\Delta p = 0$, $\Delta x = \infty$, a very small wave packet ('particle' with fixed position) has $\Delta x = 0$, $\Delta p = \infty$. In Chapter IV, section 1, we shall find a mathematical criterion which will enable us to tell which quantities have a sharp value in a certain instance.

Our next task will be to derive a general equation for the propagation of the wave amplitude or wave function ψ which is to replace Newton's equation of motion for a particle. Before doing so, some further observations about the uncertainty principle must be made.

The right-hand side of (8) is also proportional to $1/m$ (if v is used, not p). It follows, then, that for a very heavy particle for which h/m is very small, the product of the two uncertainties $\Delta x \Delta v$ becomes very small too. In this case both the position and the velocity are practically sharp.† Especially for macroscopical bodies with m so large that $h/m \simeq 0$, all uncertainties vanish, and all quantities have sharp values. This is the limiting case of *classical mechanics*. Classical mechanics holds for heavy bodies; the 'uncertainties' are a peculiarity of quantum

† The momentum, though, will still be unsharp, but for a heavy body a moderately small momentum means vanishing velocity.

mechanics which applies to light particles, especially electrons, protons, etc.

A further question which will present itself if the foregoing principles are to be logical and not self-contradictory is the following: we have stated that in the diffraction experiment the momentum has a sharp value (it was determined by experiment) whilst the position was uncertain. But what happens when the electron has arrived on the screen and the scintillation is *observed* at a particular point? Then we obviously *do know* the position of the electron and we could hardly admit that the latter is still uncertain.

We may ask, for example, what would happen if we place a second scintillation screen immediately behind the first (making the first screen so thin that the electron can pass through) and observe the position once more on the

FIG. 5. Two subsequent observations of position of electron on two scintillation screens.

second screen. If there is any meaning at all in saying 'the electron was found at a certain point' then surely we must find it again at the corresponding point on the second screen (Fig. 5), otherwise the first statement of position would be valueless. On the other hand, what has now become of the wave function after the position is observed? ψ^2 is the probability distribution of the position and this extends over a large area (Fig. 4). If ψ were to remain unchanged there would be no reason whatsoever why the electron should appear at the same point on the second screen. It might jump to another point with an equally large probability, perhaps a few centimetres away! But this is not what happens.

The answer to this question is as simple as it is far-reaching. We state the answer without being able to develop the theory in this book up to a point where it would appear as part of the general structure of quantum mechanics.

The moment the electron has been observed at a particular point on the screen it has, indeed, a sharp position. Then the electron is to be described by a small wave packet which has a sharp position but an uncertain momentum. Accordingly, the wave function ψ changes from a monochromatic wave, *suddenly* and decisively, into a small wave packet. The change is effected by the *measurement* of the position on the screen. By this measurement we *force* the electron into a state with sharp position. (A subsequent second measurement of position, following immediately after the first, yields then the same result.) At the same time we *destroy the knowledge* we had before of the momentum (gained by measurement!). If, afterwards, the momentum were measured again we should find all sorts of values for p, each with a certain probability, and these values would be very different from the momentum the electron had when it left the cathode-ray tube. It depends on the experiment performed which quantities have sharp values. A quantity that is *measured* is *forced* to be *sharp*.

In atomic physics a measurement no longer leaves the measured object uninfluenced. Two measurements of the same quantity (p, say) carried out in succession will not yield the same result if a different quantity (x, say) is measured in between them. This is a very striking feature of wave mechanics and contrary to classical physics.

The sudden change of the ψ function shows that the wave field of an electron is something very different from the familiar kind of field such as the electric or magnetic field. The latter is directly observable and measurable, it never changes suddenly, and a measurement has no influence on it. The wave field of an electron is nothing that is observable directly. Its physical meaning is a probability (rather a 'probability amplitude', for the probability is ψ^2). For that reason it is nevertheless an essential ingredient of the nature of the electron and vital for its description. For without the ψ function no laws of nature could be formulated in the domain of atomic physics.†

These considerations were given for the sake of logical completeness. They will not be further needed in this book. In atomic physics one is nearly always concerned with cases where the energy is sharp. We shall develop the theory of wave mechanics mainly for this case.

† The philosophically minded reader may decide for himself whether he would consider the wave field of an electron (ψ function) as part of an 'objective reality' or 'only' as a product of the human brain, useful for predicting the results of experiments (it happens, though, that these predictions always agree with the 'objective' facts). The author does not wish to influence him in his belief on this point. He only suggests that the question of what 'objective reality' is, be cleared up first.

II

DERIVATION OF THE WAVE EQUATION

1. The free electron

A MONOCHROMATIC wave travelling in the direction of the x-axis, with wave-length λ and frequency ν, has an amplitude ψ given by the equation

$$\psi = A \sin 2\pi\left(\frac{x}{\lambda} - \nu t\right) \tag{1}$$

or by the corresponding cosine expression, which describes the same wave as (1) with a shift of phase. There exist also standing waves, in which the maxima and minima do not travel along:

$$\psi = A \sin \frac{2\pi x}{\lambda} \cos 2\pi\nu t. \tag{2}$$

The sin and cos can here also be replaced by cos and sin respectively.

The standing waves are more important for describing the motion of an electron in an atom. In a discrete level we certainly are not dealing with a wave running to infinity but it will be seen that the electron behaves as such a standing wave. Hence we will confine ourselves to waves depending on time like (2). Furthermore, we shall consider the space part, $A \sin 2\pi x/\lambda$, and the time part, $\cos 2\pi\nu t$, separately. We shall see that for the type of problem considered below we need not be concerned with the latter.

18 DERIVATION OF THE WAVE EQUATION

Hence we may use either of the two expressions:

$$\left.\begin{aligned}\psi &= A\sin\frac{2\pi x}{\lambda} \\ \psi &= A\cos\frac{2\pi x}{\lambda}\end{aligned}\right\}, \qquad (3)$$

or, more generally,

$$\psi = A\cos\left(\frac{2\pi x}{\lambda}+\delta\right) \qquad (4)$$

where δ is an arbitrary phase between 0 and 2π.

In wave mechanics one often has to deal also with complex waves containing the imaginary unit $i = \sqrt{(-1)}$. Instead of (3) we may, for instance, also write

$$\psi = A\left(\cos\frac{2\pi x}{\lambda}+i\sin\frac{2\pi x}{\lambda}\right) = Ae^{2\pi ix/\lambda}. \qquad (5)$$

This is also a wave with wave-length λ. In particular we shall see (section 4) that the time-dependent factor is *always*
$$e^{-2\pi i\nu t} = \cos 2\pi\nu t - i\sin 2\pi\nu t \qquad (6)$$

instead of sin or cos. Differentiating (4) or (5) twice we find

$$\frac{d^2\psi}{dx^2} = -\left(\frac{2\pi}{\lambda}\right)^2\psi.$$

The waves satisfy the equation

$$\frac{d^2\psi}{dx^2}+\left(\frac{2\pi}{\lambda}\right)^2\psi = 0. \qquad (7)$$

This is a characteristic 'wave equation' and is the basis of the whole theory.

We now wish to relate (7) to the motion of an electron considered as a particle. In order to do this we express

DERIVATION OF THE WAVE EQUATION

the wave-length λ by the particle velocity v using the relation $\lambda = h/mv$. The energy is then

$$E = \tfrac{1}{2}mv^2 = \tfrac{1}{2}m\left(\frac{h}{m\lambda}\right)^2.$$

Hence the expression $(2\pi/\lambda)^2$ occurring in the above wave equation (7) becomes

$$\left(\frac{2\pi}{\lambda}\right)^2 = \frac{8\pi^2 mE}{h^2}. \tag{8}$$

We thus obtain a wave equation in which the energy E of the particle occurs. For convenience we introduce instead of h, $h/2\pi = \hbar$. Hence the wave equation becomes

$$\frac{d^2\psi}{dx^2} + \frac{2mE}{\hbar^2}\psi = 0. \tag{9}$$

Every solution of (9) is to be multiplied by a factor $\exp(-2\pi i\nu t)$ (using the complex expression (6)); ν is directly connected with the energy E occurring in (9): $2\pi\nu = E/\hbar$. In the following the time-dependent factor will not be of much interest.

So far the wave equation (9) describes only waves extending in one dimension, namely, in the x-direction. Naturally, ψ will, in general, depend on all three coordinates x, y, z. If we had, for instance, a wave extending in the y-direction $d^2\psi/dx^2$ would have to be replaced by $d^2\psi/dy^2$. It is not difficult to see that in the general case $d^2\psi/dx^2$ is to be replaced by

$$\frac{\partial^2\psi}{\partial x^2} + \frac{\partial^2\psi}{\partial y^2} + \frac{\partial^2\psi}{\partial z^2}.$$

DERIVATION OF THE WAVE EQUATION

For this expression we use the short notation $\nabla^2 \psi$. Hence the general wave equation is

$$\nabla^2 \psi + \frac{2mE}{\hbar^2} \psi = 0, \qquad (10)$$

where $\nabla^2 = \dfrac{\partial^2}{\partial x^2} + \dfrac{\partial^2}{\partial y^2} + \dfrac{\partial^2}{\partial z^2}.$

The wave equation (10) describes monochromatic waves only, with a particular given frequency ν or energy E. E occurs in (10) directly and also in the time-dependent factor (6). For all solutions of (10) the energy has therefore a sharp value. Naturally, there may also be waves for which E is not sharp (for instance wave packets) and whose wave function will not then be proportional to a factor (6) with one given frequency ν, but will depend on time in a more complicated way, allowing many frequencies, i.e. different energies, to occur. In order to form a wave packet, for instance, we must take several different solutions of (10) with different E's, multiply each by its appropriate time-factor, and take a linear combination of them. (Similar to the representation of a sound pulse by monochromatic waves.) Such a wave packet is no longer a solution of (10) but is a solution of a more general wave equation, the so-called time-dependent wave equation. This will be derived and discussed in section 4. Otherwise, we shall throughout this book only be interested in cases for which E has a sharp value, i.e. in monochromatic waves. In this case (10) is the wave equation always to be used. The time-factor is then always (6) and of no interest. It will henceforth be omitted.

The wave equation (10) and the more general equation (17), derived in section 3, replace *Newton's equation of*

DERIVATION OF THE WAVE EQUATION

motion of a particle. Whilst Newton's equation of motion allows one to calculate the orbit of a particle for any time accurately (in any given circumstances), the wave equation only allows one to calculate ψ, i.e. the probability for finding the particle at a certain position. This replacement of accurate and predictable orbits by uncertain values and probability distributions is the chief step in the transition from classical physics to wave mechanics. It is due, as we have seen, to the double nature of electrons, as particles and waves.

2. Discrete quantum states

From the wave equation (10) we can draw a most important conclusion, namely, the existence of *discrete energy levels*. We have stated in Chapter I that an atom can only exist in a number of discrete energy states. We now derive this important result for a very simple case. We consider an electron moving along the x-axis, between two reflecting walls which are perfect mirrors for the electron. Considered from the wave point of view this is a case where we have to deal with standing waves. It is similar to the vibration of a violin string fixed at two end-points. Since the electron cannot pass through the walls the amplitude ψ must be zero outside the walls, and therefore at each wall itself. The amplitude, ψ, is therefore zero at $x = 0$ and L, say. These are new additional conditions which we call 'boundary conditions'. The standing waves must therefore have a node at each wall. Hence it follows that not all wave-lengths are possible, but only those, as we see from Fig. 6, which have wave-length $\lambda = 2L$ or $2L/2$, $2L/3$, $2L/4$, etc. This, for instance, is the condition determining the tone of a violin string. $\lambda = 2L$ gives the ground tone and the other wave-lengths $\lambda = 2L/2, 2L/3,...$ give the different overtones.

DERIVATION OF THE WAVE EQUATION

In mathematical form this is derived as follows: The solution of the wave equation is either of the two forms

$$\left.\begin{aligned}\psi &= A \sin \frac{2\pi}{\lambda} x \\ &= A \cos \frac{2\pi}{\lambda} x\end{aligned}\right\} . \tag{11}$$

or

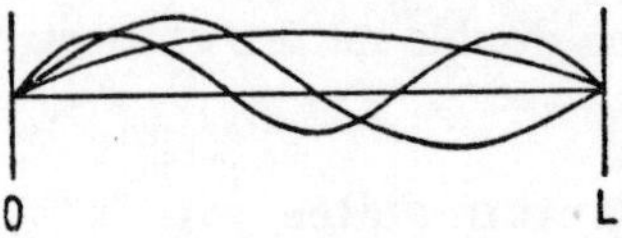

FIG. 6. Discrete wave functions (electron between two walls).

The cosine solution does not satisfy the boundary condition $\psi = 0$ at $x = 0$, and must therefore be discarded. The sine solution satisfies this condition. The second boundary condition is $\sin \frac{2\pi}{\lambda} L = 0$, and this is only true for certain values of λ, namely:

$$\frac{2\pi}{\lambda} L = n\pi, \tag{12}$$

where n is a whole number.

The latter condition gives the value of λ straight away as

$$\lambda_n = \frac{2L}{n}, \quad \lambda_1 = 2L, \quad \lambda_2 = \frac{2L}{2}, \quad ..., \tag{13}$$

where $n = 1, 2, 3, 4,...$, etc. There is an infinite number of solutions, since for each value of n we obtain a wavelength, say, λ_n and a corresponding solution

$$\psi_n = A_n \sin \frac{\pi n}{L} x. \tag{14}$$

The amplitudes A_n may, of course, be different for each n.

DERIVATION OF THE WAVE EQUATION

For the purpose of showing the discreteness of the energy levels we introduce instead of λ the energy

$$E = \frac{m}{2} v^2 = \frac{m}{2}\left(\frac{h}{m\lambda}\right)^2.$$

Then, inserting for λ the possible values of λ_n from (13), we have

$$E_n = \frac{m}{2} \frac{h^2}{m^2} \frac{n^2}{4L^2} = \frac{h^2 n^2}{8L^2 m}. \tag{15}$$

Therefore only *certain values of E*, namely, E_n, *are possible* and these are the only ones allowed. The lowest energy is

$$E_1 = \frac{h^2}{8mL^2} \quad (n = 1), \tag{16}$$

and the higher energies are 4, 9,... times this value. There are no energy values in between these values. The energy values E_n are usually called *eigenvalues* and ψ_n *eigenfunctions*.

If these discrete energy levels are plotted (Fig. 1) it is seen that the spacing increases as $2n+1$. It is due to the boundary conditions that the energy levels are discrete. Our example shows how the discrete quantum states can be understood from the wave-mechanical viewpoint.

In an atom, however, the position is not quite the same as in our example, since in an atom there are no actual boundaries. On the other hand, the electron is moving in the field of an electric potential given by $V = -e^2/r$ shown in Fig. 7. Now if the electron is bound to the nucleus, its energy must be negative. On the other hand, its kinetic energy is always positive. Therefore, if we consider the motion of the electron classically, it can only move up to a maximum value of r shown by the curves of Fig. 7; the kinetic energy is zero when the electron touches the curves. This is indeed a condition similar to the boundary conditions considered above.

In wave mechanics the boundary condition is slightly different because the curves of Fig. 7 are not really impenetrable walls with an infinitely high potential as in the example of Fig. 6. Nevertheless, if the electron has negative energy and is therefore bound, it cannot leave the atom.

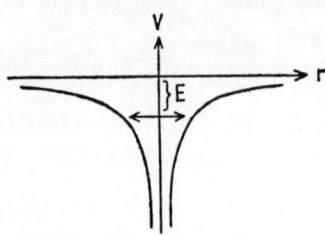

Fig. 7. Electron in Coulomb potential.

This means that ψ must be zero at large distances and this condition is quite sufficient to lead to discrete energy levels of the atom (see Chapter III, especially the end of section 1) but, of course, the spacing of the levels is quite different from that in the above example.

The fact that, in classical theory, the electron could not move beyond the curves of Fig. 7, finds its expression in wave mechanics in that the ψ-function decreases rapidly to zero for values of r larger than the classical maximum.

3. The Schrödinger wave equation

The wave equation derived in section 1 applies still to electrons moving freely in space but not subject to any external forces. If an electron moves under the influence of an electric field, say, it is deflected and accelerated and consequently we must expect that in the wave picture also the waves are somehow deflected. We must therefore generalize our wave equation to take account of such external forces. Let us consider an electrical potential and try to obtain a wave equation which shall be general for such cases. The equation describing the free motion of an electron makes no distinction between kinetic energy, T, and total energy, E. The total energy is $E = T+V$, where V is the potential energy. If, therefore, the electron

DERIVATION OF THE WAVE EQUATION

is not moving in an external field, $V = 0$ and the energy E in the wave equation (10) is equal to the total energy and the kinetic energy at the same time. But if a potential V is present, the question arises whether we should then insert in the wave equation, instead of E, the total energy $E = T+V$, or the kinetic energy, $T = E-V$. If we should decide for the former alternative the wave equation would remain unaltered, E being then the total energy. The energy levels would all be the same as for a free electron and the wave function ψ also the same, whatever the potential V might be. This certainly cannot be true. We therefore assume that E in the wave equation for the free electron means, in fact, the kinetic energy, i.e. $E-V$, if the electron moves in a field of potential V. The general wave equation then becomes

$$\nabla^2 \psi + \frac{2m}{\hbar^2}(E-V)\psi = 0. \qquad (17)$$

(17) is the celebrated wave equation found by Schrödinger (1926). It describes quite generally the motion of an electron under the influence of any electrical potential V.

Our procedure, namely the replacing of E by $T = E-V$, is also supported by the following purely formal considerations, which will also be useful later on.

Let us rewrite the wave equation (10) for a free particle in the form

$$E\psi = -\frac{\hbar^2}{2m}\nabla^2\psi = -\frac{\hbar^2}{2m}\left(\frac{\partial^2}{\partial x^2}+\frac{\partial^2}{\partial y^2}+\frac{\partial^2}{\partial z^2}\right)\psi, \qquad (18)$$

and let us compare this equation with the corresponding expression for E in classical theory:

$$E = \frac{m}{2}(v_x^2+v_y^2+v_z^2).$$

Introducing $mv_x = p_x$ as the momentum, we have

$$E = \frac{1}{2m}(p_x^2 + p_y^2 + p_z^2). \tag{19}$$

We see that a great similarity exists between the two equations. In order to pass from the classical expression for E, (19), to the wave equation, (18), we have to substitute for p_x^2 the 'operator' $-\hbar^2(\partial^2/\partial x^2)$, and let this operator act on the wave function ψ. Or, in other words, we have to replace p_x itself by the operator:

$$p_x \longrightarrow \frac{\hbar}{i}\frac{\partial}{\partial x}. \tag{20}$$

This replacement seems mysterious at this stage, but is part of the general theory of quantum mechanics, where it is seen to have a deep meaning which cannot be fully explained in this book.

If we now consider the case of an electron moving in a potential V, the classical equation for the energy is

$$T = E - V = \frac{1}{2m}(p_x^2 + p_y^2 + p_z^2).$$

If we again replace each p_x by the operator $\frac{\hbar}{i}\frac{\partial}{\partial x}$ and let it operate on a wave function ψ, we obtain $T\psi = (E-V)\psi$, which is just the wave equation (17). Ultimately, it is of course experiment which has decided in favour of the wave equation (17) and therefore also the formal replacement (20).

Thus the wave equation (17) bears a close relation to the classical equation for the energy. If we denote the classical energy function expressed as a function of the coordinates and momenta by $H(x,...,p_x,...)$, also called 'Hamiltonian',

$$E = H(x,...,p_x,...) = \frac{1}{2m}(p_x^2 + p_y^2 + p_z^2) + V(x,y,z),$$

$$(21\,a)$$

DERIVATION OF THE WAVE EQUATION

the wave equation is

$$E\psi = H\psi, \qquad H \equiv H\left(x,...,\frac{\hbar}{i}\frac{\partial}{\partial x},...\right). \qquad (21\,b)$$

Again, all the solutions of the wave equation (17) are those with sharp energy. This does not mean, and this is in contrast to the case of a free electron, that the momentum is also sharp. For now the energy E is the sum of the kinetic energy T and the potential energy V. The latter depends on the coordinates of the electron which certainly are not sharp. Therefore, if $E = T+V$ is to be sharp, T or p cannot be sharp (see Chapter III). This is connected with the fact that in classical theory the momentum of the electron does not remain constant in course of time, except for a free electron ($V = 0$). This finds its expression in quantum theory (as we cannot explain here) in the fact that when the energy is sharp only such quantities can also be sharp (but need not necessarily be so) which classically remain constant in course of time. An example (angular momentum) is considered in Chapter IV. For the various values of p a probability distribution, analogous to the ψ^2 for the position, exists, but this will not be needed in this book.

4. The time-dependent wave equation

The wave equation (17) only applies to cases where the energy is sharp. We have already stated that this is not the most general case and that in fact a time-dependent wave equation exists, which also describes running waves and has solutions for which E is not sharp. Also, even if E is sharp, the complete wave function is not merely the solution of (17), $\psi(x,y,z)$, but this is to be multiplied by a time-dependent factor like $\exp(-iEt/\hbar)$. Now we seek a more general equation which does not contain the energy E explicitly but which reduces to (17), multiplied by a time-factor, as a special case for sharp E. In general, the solution of such a wave equation will be a function $\psi(x,y,z,t)$ of the coordinates as well as of the time. Now a wave equation which such a function is to satisfy must only contain the derivatives of ψ with respect to the coordinates as well as the time but must not contain any constant like the energy or the wave-length. Such constants occur in special solutions only. On the other hand, the potential V will occur because this describes the external conditions under which the electron moves.

Thus it is clear that the term $(2mE/\hbar^2)\psi$ which occurs in (17) will be replaced by some derivative of $\psi(x,...,t)$ with respect to the time. The question is now whether this is the first derivative $\partial\psi/\partial t$ or perhaps the second derivative $\partial^2\psi/\partial t^2$ as in the case of the coordinates ($\partial^2\psi/\partial x^2$ etc.). This is easily settled if we remember that a special solution must be (17) multiplied by a time-factor

$$\exp(-iEt/\hbar).$$

Since in (17) E and not E^2 occurs we see that only the first derivative can occur. Thus instead of the term $\sim E$, we shall have a term $\sim \partial\psi/\partial t$. In order to obtain the correct factors in (17) we must then write the wave equation as

$$\nabla^2\psi - \frac{2m}{\hbar^2}V\psi + \frac{2im}{\hbar}\frac{\partial\psi}{\partial t} = 0. \qquad (22)$$

Indeed, we obtain a solution with sharp energy E if we put

$$\psi(x,y,z,t) = \psi(x,y,z)e^{-iEt/\hbar} \qquad (22')$$

and then $\psi(x,y,z)$ satisfies the old equation (17). (22) is now the general time-dependent wave equation. When solved, it tells us quite generally how the wave function develops in course of time. For sharp energy E we see that the correct time-factor is the exponential and not

DERIVATION OF THE WAVE EQUATION

$\cos Et/\hbar$. The latter would not lead back to (17) because the first derivative of cos (or sin) does not reproduce this function. It is remarkable that in the wave equation (22) the imaginary unit i occurs.† In general ψ will therefore be complex and this again underlines the fact that ψ cannot be a measurable field quantity like the electrical field strength.

Previously we had interpreted ψ^2 as the probability for finding the electron at a certain point. Now when ψ is complex it need hardly be proved that this probability will be $|\psi|^2$. Most of the solutions occurring in this book, however, will be real and in this case the probability is ψ^2.

The solution of (22) which describes a running wave for $V = 0$ is obviously

$$\psi = A e^{2\pi i x/\lambda - iEt/\hbar} = A e^{2\pi i (x/\lambda - \nu t)} \qquad (23)$$

with $$E = \frac{h^2}{2m\lambda^2} = \frac{m}{2} v^2.$$

(Again it is not the real form $\cos 2\pi(x/\lambda - \nu t)$.) (23) is a wave travelling along the positive x-axis with wave-length λ and frequency ν or phase velocity $\lambda\nu$.

A still more general solution is obtained by superposing several running waves of the type (23) with different frequencies. As a simple example we superpose two waves with but slightly different wave-lengths λ, say $\lambda = \lambda_0 + \Delta\lambda$ and $\lambda = \lambda_0 - \Delta\lambda$ where $\Delta\lambda$ is small compared with λ_0. This example will afford a derivation and illustration of the group velocity v_g. The frequencies are $\nu_0 + \Delta\nu$ and $\nu_0 - \Delta\nu$. Let the amplitudes of both waves be the same. Using then the fact that

$$\frac{1}{\lambda_0 + \Delta\lambda} \simeq \frac{1}{\lambda_0} - \frac{\Delta\lambda}{\lambda_0^2},$$

---

† The sign with which i occurs in (22) as well as in the time-factor (22′) is irrelevant and a mere convention as long as the same sign is used consistently.

the solution is

$$\psi = A\{e^{2\pi i[x/(\lambda_0+\Delta\lambda)-(\nu_0+\Delta\nu)t]} + e^{2\pi i[x/(\lambda_0-\Delta\lambda)-(\nu_0-\Delta\nu)t]}\}$$

$$= 2A e^{2\pi i(x/\lambda_0-\nu_0 t)} \cos 2\pi\left(\frac{\Delta\lambda}{\lambda_0^2}x + \Delta\nu t\right). \quad (24)$$

It may be left to the reader to show that (24) is really a solution of (22). Now there is a relation between ν and λ, $\nu = h/2m\lambda^2$ and therefore, upon differentiation,

$$\Delta\lambda = -\frac{m\lambda_0^3}{h}\Delta\nu, \quad (25)$$

where we have inserted λ_0 for λ. Thus the wave function is

$$\psi = 2A e^{2\pi i(x/\lambda_0-\nu_0 t)} \cos 2\pi \Delta\nu\left(\frac{m\lambda_0}{h}x - t\right). \quad (26)$$

The first factor (the exponential) describes an ordinary running wave with wave-length λ_0. However, this wave is now 'modulated' by the second factor (the cos). For a

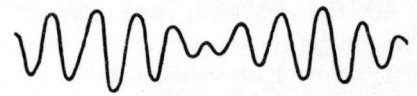

FIG. 8. Modulated wave.

fixed time $t = 0$, for example, this is a periodic factor and has maxima at $x = 0$, $x = h/m\lambda_0\Delta\nu$, etc. Since $\Delta\nu$ is small, this is a wave with a very *long wave-length* $h/m\lambda_0\Delta\nu$. The whole modulated wave looks like Fig. 8. Now as a function of $t \neq 0$ and of x the modulating factor also describes a wave travelling along the x-axis. The maxima of the modulating factor, however, travel with a velocity, given by the ratio of the factors of x and t in the cosine of (26). We call this velocity v_g. It is given by

$$v_g = \frac{h}{m\lambda_0} = v. \quad (27)$$

DERIVATION OF THE WAVE EQUATION

The right-hand side is the particle velocity v (p. 4, eqn. (2)) which we had identified with the 'group velocity' v_g (compare Fig. 3). The latter was defined as $v_g = d\nu/d(1/\lambda)$ and since $\nu = E/h = p^2/2mh = h/2m\lambda^2$, we obtain

$$v_g = (h/2m)\frac{d(1/\lambda^2)}{d(1/\lambda)} = h/m\lambda = v.$$

Thus the superposition of two waves with slightly different wave-lengths provides a simple example for the group velocity and its identity with the particle velocity. For the rest of this book the time-dependent wave equation will not be used any more.

III

THE HYDROGEN ATOM

1. The ground state

In the H atom we have one electron with a charge $-e$ rotating round a central proton P with charge $+e$. We take the proton as the centre of our system of coordinates. The potential is then $V = -e^2/r$, and substituting this in the wave equation II (17), we get

$$\nabla^2 \psi + \frac{2m}{\hbar^2}\left(E + \frac{e^2}{r}\right) = 0. \qquad (1)$$

We shall not attempt to find all the solutions of this equation, which would require some mathematical labour. Instead we shall give a few simple solutions.

The potential e^2/r is spherically symmetrical. There will be a class of solutions ψ which are also spherically symmetrical, i.e. for which ψ depends on r only. To solve (1) let us remember that $r = \sqrt{(x^2+y^2+z^2)}$ and therefore that

$$\frac{\partial r}{\partial x} = \frac{1}{2}\frac{2x}{\sqrt{(x^2+y^2+z^2)}} = \frac{x}{r}.$$

Consider now

$$\nabla^2 \psi = \left(\frac{\partial^2}{\partial x^2} + \frac{\partial^2}{\partial y^2} + \frac{\partial^2}{\partial z^2}\right)\psi.$$

If ψ depends only on r but not on the polar angles,

$$\frac{\partial \psi}{\partial x} = \frac{\partial \psi}{\partial r}\frac{\partial r}{\partial x}$$

or
$$\frac{\partial \psi}{\partial x} = \frac{x}{r}\frac{\partial \psi}{\partial r},$$

THE HYDROGEN ATOM

and differentiating again,

$$\frac{\partial^2 \psi}{\partial x^2} = \frac{1}{r}\frac{\partial \psi}{\partial r} - \frac{x^2}{r^3}\frac{\partial \psi}{\partial r} + \frac{x^2}{r^2}\frac{\partial^2 \psi}{\partial r^2}.$$

For the remaining terms we simply replace x by y and z:

$$\frac{\partial^2 \psi}{\partial y^2} = \frac{1}{r}\frac{\partial \psi}{\partial r} - \frac{y^2}{r^3}\frac{\partial \psi}{\partial r} + \frac{y^2}{r^2}\frac{\partial^2 \psi}{\partial r^2}$$

and

$$\frac{\partial^2 \psi}{\partial z^2} = \frac{1}{r}\frac{\partial \psi}{\partial r} - \frac{z^2}{r^3}\frac{\partial \psi}{\partial r} + \frac{z^2}{r^2}\frac{\partial^2 \psi}{\partial r^2}.$$

Adding these three second differentials together and remembering that $x^2 + y^2 + z^2 = r^2$, we find

$$\nabla^2 \psi = \frac{3}{r}\frac{\partial \psi}{\partial r} - \frac{1}{r}\frac{\partial \psi}{\partial r} + \frac{\partial^2 \psi}{\partial r^2} = \frac{\partial^2 \psi}{\partial r^2} + \frac{2}{r}\frac{\partial \psi}{\partial r}.$$

The wave equation then becomes

$$\frac{\partial^2 \psi}{\partial r^2} + \frac{2}{r}\frac{\partial \psi}{\partial r} + \frac{2m}{\hbar^2}\left(E + \frac{e^2}{r}\right)\psi = 0. \tag{2}$$

The simplest solution of (2) is

$$\psi(r) = e^{-ra}, \tag{3}$$

where a is a constant to be determined.

Substituting (3) into (2), we have

$$\frac{\partial \psi}{\partial r} = -ae^{-ra},$$

$$\frac{\partial^2 \psi}{\partial r^2} = +a^2 e^{-ra}.$$

It will be noticed that e^{-ra} occurs in each term of (2). Therefore, dividing by this factor throughout, we obtain

$$a^2 - \frac{2}{r}a + \frac{2m}{\hbar^2}\left(E + \frac{e^2}{r}\right) = 0. \tag{4}$$

Now collecting the terms which are independent of r in this equation, namely, the first and third terms, and then those proportional to $1/r$, namely, the second and fourth terms, we see that both must be equal to zero, because (4) holds for all values of r.

$$\left.\begin{array}{c}\dfrac{2mE}{\hbar^2}+a^2=0\\[6pt]a=\dfrac{me^2}{\hbar^2}\end{array}\right\}, \qquad (5)$$

and hence $\qquad E = -\dfrac{a^2\hbar^2}{2m} = -\dfrac{me^4}{2\hbar^2}. \qquad (6)$

If E and a have the above values, the suggested exponential function $\psi(r) = e^{-ra}$, (3), is a solution. It also satisfies the boundary conditions described in Chapter II, section 2, since (3) becomes zero at large distances from the nucleus. Actually, this is the solution with the smallest value of E (which cannot, however, be shown here), and therefore describes the ground state of the H atom.

First we note that E is negative, therefore the electron is bound. E is determined by m, e, and $\hbar$ entirely and has, therefore, a fixed value, since these are universal constants. It can be measured experimentally. $-E$ is the 'ionization energy', that is, the energy required to remove the electron from the nucleus. This energy is known from spectroscopy to be equal to 13·5 eV (electron volts). On the other hand, the values of the universal constants m, e, $\hbar$ are also well known. Inserting these values into the expression $E = -me^4/2\hbar^2$, we find also $-E = 13\cdot5$ eV which is in perfect agreement with the experimental value.

We now consider the solution $\psi(r)$ itself. $\psi^2 d\tau$ gives the probability of finding the electron in the volume element $d\tau$. If we introduce polar coordinates, $4\pi r^2 dr$ is the volume

THE HYDROGEN ATOM

of a shell with radius r, and the probability of the electron being in this shell is therefore proportional to $r^2\psi^2$. Plotting this function $r^2 e^{-2ra}$ against r (Fig. 9) we observe that the probability of finding the electron at a distance r from the nucleus is small at small distances (because of the factor r^2) and at very large distances (because of the factor

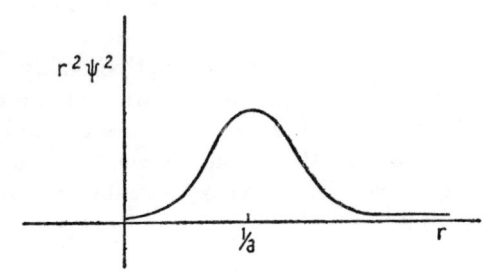

FIG. 9. Probability distribution of hydrogen ground state.

e^{-ra}). It has a maximum at $r = 1/a$. The distance $1/a$ is the most probable distance of the electron from the nucleus. The length $1/a$ is also expressed by the universal constants, m, e, $\hbar$ (equation (5)), and is called the 'Bohr radius'. It first occurred in Bohr's theory of the H atom, in which the electron was assumed to move on a circle with radius $1/a$ round the nucleus. It is interesting to note that such an electron would have a potential energy $-e^2 a = 2E$, by (6). The amount $+me^4/2\hbar^2 = -E$ is the kinetic energy giving a total energy E. Although Bohr's theory was the first to give an account of the quantum phenomena in an atom, it is now replaced by the more consistent and more comprehensive theory of quantum mechanics (or wave mechanics). The numerical value of the Bohr radius, as found from (5), is $1/a = 0{\cdot}53 \times 10^{-8}$ cm.

We know now that the electron does not quite move on a circle, as its position is not, for any time, certain, but we can say that the probability for finding the electron at a certain distance from the nucleus is biggest if that distance is equal to the 'Bohr radius'.

The probability distribution, Fig. 9, affords an illustration of the uncertainty principle. While in classical theory (and in Bohr's theory) the kinetic energy of an electron in a Coulomb field is $T = -E$, this cannot be true in wave mechanics because neither V nor T is sharp. Nevertheless, this relation will hold approximately, in the following sense. From Fig. 9 it appears that the position of the electron is uncertain within limits of the order of magnitude $\Delta x = 1/a$. Hence the momentum is uncertain to within $\Delta p = \hbar a$. The average absolute value of the momentum must be at least of the same order of magnitude. In the ground state the electron will not have more kinetic energy or momentum than it need have. Thus $\hbar a$ will actually be the average momentum. Consequently, the average kinetic energy will be

$$\overline{T} = \frac{p^2}{2m} = \frac{1}{2m}\hbar^2 a^2 = \frac{me^4}{2\hbar^2}.$$

This is just equal to $-E$, i.e. the same as in Bohr's theory.

Another interesting feature of our solution e^{-ra} is the following: Although ψ decreases rapidly for large r, it is not quite zero, however large r may be. There is therefore a small but finite probability for finding the electron even at large distances from the nucleus. In classical theory this could never happen. For there the kinetic energy can never be negative and the particle can never reach a point where the potential energy would be larger than its own total energy E. The potential energy being $V = -e^2/r$ and the total energy E given by (6), we see that r could not be larger than $2\hbar^2/me^2 = 2/a$. Such a potential barrier as is represented by V (see Fig. 7) cannot be penetrated by a particle, according to classical mechanics. This is not so in wave mechanics. On account of its wave nature the electron has a certain small chance of penetrating such a potential barrier and of being found outside it (unless the barrier is infinitely high as in the example (Chapter II) of an electron moving between two walls). This is in fact one of the most conspicuous consequences of wave mechanics. At first sight this also seems to contradict the principle of conservation of energy, for the electron does not have the energy to move beyond the distance $2/a$. Our statement is to be regarded in the following sense. If we wish to observe the position we have to carry out an experiment. Now such an experiment has a decisive influence on the electron itself (it changes its wave function (Chapter I)). It can now be shown that the measurement also supplies the necessary energy. If the electron is observed at a large distance from the proton the energy required to move it there is supplied by the measuring apparatus.

2. Excited states

In addition to the state of lowest energy there are also higher states with energies $E_n > E_1$, if we now denote the energy of the ground state given by (6) by E_1. We shall not derive all these solutions here but only give the results. The energies of the various excited states are:

$$E_2 = -\frac{me^4}{2\hbar^2}\frac{1}{2^2} = \frac{1}{4}E_1, \tag{7}$$

or generally, $\quad E_n = -\dfrac{me^4}{2\hbar^2}\dfrac{1}{n^2} = \dfrac{1}{n^2}E_1, \tag{7'}$

where $n = 1, 2, 3,..., \infty$.

If n is very large, E is very small, and in the limit $n \to \infty$ E_n becomes zero. This is then the case of a *free* electron.

The energies (7) agree also with the experimental facts: Balmer found that the frequency of the spectral lines emitted by hydrogen can be represented by the formula

$$\nu = \left(\frac{1}{m^2} - \frac{1}{n^2}\right) \times \text{const.} \tag{8}$$

According to Chapter I the frequencies of such spectral lines are given by

$$h\nu = E_n - E_m.$$

If we insert (7) for E_n we just obtain the Balmer formula (8).

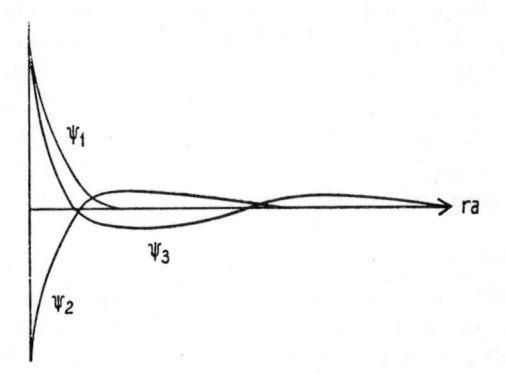

Fig. 10. Wave functions of hydrogen.

The wave functions ψ_n belonging to the energies E_n (7) are also spherically symmetrical, i.e. functions of r only, or more precisely: To each energy value (7) there is a spherically symmetrical solution $\psi_n(r)$. We leave it to the reader to verify, for instance, that $\psi_2 = e^{-ra/2}(2-ra)$ is a solution of (2) belonging to the energy value E_2 (7). This solution, together with ψ_1 and ψ_3 belonging to the energies E_1 and E_3, are shown in Fig. 10 (multiplied by suitable factors). As we may well expect, the behaviour of these

THE HYDROGEN ATOM

solutions is somewhat similar to those of an electron moving between two walls (Fig. 6). All ψ_n's decrease, however, rapidly for large r and satisfy the boundary conditions, namely, that ψ should vanish for large r. The average distance of the electron from the nucleus is increasingly larger for the higher ψ_n's than for ψ_1.

States whose wave functions are spherically symmetrical are called s states. There are, however, also solutions whose wave functions are *not* spherically symmetrical. These we shall study in section 3.

Apart from the states with negative energy which form what is called a discrete spectrum E_n (7') there is also an infinite variety of solutions with positive energy. These describe a free electron moving in the field of the nucleus and correspond to the hyperbolic orbits of classical theory. The electron comes from and goes to infinity. Such solutions form a continuous spectrum, every value $E > 0$ is possible.

3. p states

Although the potential V in the wave equation

$$\nabla^2\psi + \frac{2m}{\hbar^2}(E-V)\psi = 0$$

depends on the distance r only, there are solutions which depend on x, y, z separately. We can show that

$$\psi = xf(r) \tag{9}$$

is a solution, if $f(r)$ is suitably determined. Inserting (9) into the wave equation we have

$$\frac{\partial \psi}{\partial x} = f + x\frac{\partial f}{\partial r}\frac{\partial r}{\partial x} = f + \frac{x^2}{r}\frac{\partial f}{\partial r},$$

$$\frac{\partial^2 \psi}{\partial x^2} = \frac{x}{r}\frac{\partial f}{\partial r} + \frac{2x}{r}\frac{\partial f}{\partial r} - \frac{x^3}{r^3}\frac{\partial f}{\partial r} + \frac{x^3}{r^2}\frac{\partial^2 f}{\partial r^2}.$$

In a similar way we find

$$\frac{\partial \psi}{\partial y} = \frac{xy}{r} \frac{\partial f}{\partial r},$$

$$\frac{\partial^2 \psi}{\partial y^2} = \frac{x}{r} \frac{\partial f}{\partial r} - \frac{xy}{r^2} \frac{y}{r} \frac{\partial f}{\partial r} + \frac{xy}{r} \frac{y}{r} \frac{\partial^2 f}{\partial r^2}.$$

$\partial^2 f/\partial z^2$ arises from $\partial^2 f/\partial y^2$ by changing y into z. Adding up the three second derivatives we obtain

$$\nabla^2 \psi = \frac{3x}{r} \frac{\partial f}{\partial r} + \frac{2x}{r} \frac{\partial f}{\partial r} - \frac{x}{r} \frac{\partial f}{\partial r} + x \frac{\partial^2 f}{\partial r^2}$$

$$= x \frac{\partial^2 f}{\partial r^2} + \frac{4x}{r} \frac{\partial f}{\partial r}.$$

If we insert this expression into the wave equation we find that each term is proportional to x. Thus, dividing by x throughout,

$$\frac{\partial^2 f}{\partial r^2} + \frac{4}{r} \frac{\partial f}{\partial r} + \frac{2m}{\hbar^2}(E-V)f = 0. \tag{10}$$

This is an equation for $f(r)$ only, which can be solved if we insert for $V(r)$ again the Coulomb potential $V = -e^2/r$ (see below). Thus we have found a solution of the type $\psi_x = xf(r)$ which is therefore not spherically symmetrical. Since, however, there is no distinction between x, y, z in the wave equation, we can say at once that

$$\psi_y = yf(r) \quad \text{and} \quad \psi_z = zf(r) \tag{11}$$

are also solutions. In (11), $f(r)$ also satisfies the same equation (10), with the same value of E. Hence there are *three solutions*, ψ_x, ψ_y, ψ_z, for each energy value of (10) with the same function $f(r)$. Any of the three functions is a solution for the same value of the energy.

A case where more than one independent solution exists for the same energy value is called *degeneracy*. In our case we are dealing with a threefold degeneracy. We may also express this fact by saying that we have three energy levels which happen to lie together. Wave functions of the type (11) are called p wave functions, and the energy state belonging to them a p state. A p state is always threefold degenerate.

The reason for the occurrence of this degeneracy is clear: It is the spherical symmetry of the potential V. Whenever we have a solution which is not spherically symmetrical itself, we can at once obtain further solutions by rotating the system of coordinates (change $x \to y$, etc.) as we have done above. This applies to *any spherically symmetrical potential*. Quite generally, symmetry properties of the potential give rise to such degeneracies. A case of axial symmetry will be discussed in Chapter VIII. Only if the wave function is itself symmetrical, is the state not degenerate. This holds also the other way round: Degeneracies do not as a rule occur unless they are caused by such symmetries of the potential V. The only exception is the case of the Coulomb field $-e^2/r$, for which an additional degeneracy occurs (see below), but this has special reasons which we cannot discuss here.

So far we have not made use of the special form of the potential $V(r) = -e^2/r$. For any spherically symmetrical potential $V(r)$ p solutions of the kind (11) exist, provided $f(r)$ satisfies (10). In the case of the Coulomb potential a solution—it is again the one with the lowest energy value—can easily be found. We put $\psi = e^{-\frac{1}{2}ar}$, with the same value of a as before, namely (5). By substituting this ψ into (10), we easily find that this is a solution, if

$$E = -\frac{me^4}{2\hbar^2}\frac{1}{4} = E_2. \tag{12}$$

This is the same energy value as that of the second lowest energy-level given by (7). Now the energy-levels (7) are those with spherically symmetrical ψ functions, i.e. of the s states. We see now that the lowest p state has the same energy as the second lowest s state. The level with energy E_2 is therefore fourfold degenerate (one s and three p wave functions). This is, however, a peculiar property of the

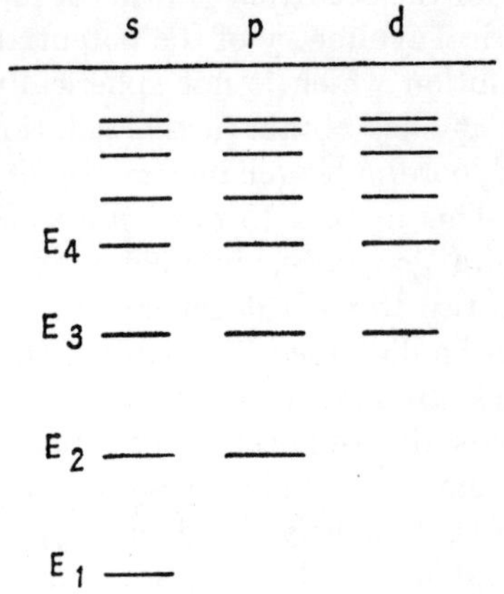

FIG. 11. Energy-levels of hydrogen.

Coulomb potential. For any other potential $V(r)$ the s and p states have different energies. For all the other atoms, for instance, the potential in which a particular electron moves is also influenced by the other electrons and therefore not simply e^2/r. The s and p states then have different energies.

There are also higher p states whose energies are also equal to E_3, E_4,..., etc. Furthermore, there are also solutions which depend in a still more complicated way on x, y, z. For instance, there are wave functions proportional to x^2, xy, xz, etc. These are called d functions and they are five-fold degenerate. Thus we obtain the level scheme, Fig. 11.

THE HYDROGEN ATOM

The p wave functions (9) and (11) may be pictured somewhat as in Fig. 12. For large values of r, ψ_x and ψ_y vanish. ψ_x is zero on the yz-plane and ψ_y on the xz-plane.

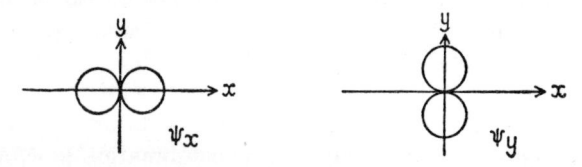

FIG. 12. p wave functions.

ψ_x is largest along the positive and negative x-axis and ψ_y along the y-axis.

In a similar way ψ_z would extend along the z-axis.

4. Normalization and linear combination of wave functions

From any solution ψ of the wave equation we can get other solutions by multiplying ψ by an arbitrary constant c; $c\psi$ is also a solution if ψ is a solution, as can be seen immediately by inserting $c\psi$ into the wave equation. What fixes the value of this constant now? Since ψ^2 is proportional to the probability of finding the electron at a certain point in space, we might fix the constant in such a way that ψ^2 is actually the absolute probability of finding the electron at a certain place. The total probability of finding the electron somewhere must then be equal to unity. Thus we have

$$\int \psi^2 \, d\tau = 1. \qquad (13)$$

This normalization condition (13) fixes the constant c, apart from the sign. A reversal of sign of ψ has no physical significance at all. The normalized solution of the ground state of hydrogen is, for instance, $a^{\frac{3}{2}}e^{-ra}/\sqrt{\pi}$, as can easily be found by integration.

Still more freedom in the choice of the functions prevails if we have two (or more) solutions belonging to the same energy. This is the case of degeneracy. Let ψ_1 and ψ_2 be two—normalized or not normalized—degenerate solutions, assuming, of course, that ψ_2 is not just a multiple of ψ_1, i.e. assuming that ψ_1 and ψ_2 are really different functions of space.

We can then get further solutions by taking the sum or difference or an arbitrary linear combination of ψ_1 and ψ_2 thus:
$$a\psi_1 + b\psi_2 = \psi.$$

ψ is also a solution if ψ_1 and ψ_2 are solutions for the same energy value

$$a\nabla^2\psi_1 + \underline{b\nabla^2\psi_2} + \frac{2m}{\hbar^2}(E-V)(a\psi_1 + \underline{b\psi_2}) = 0.$$

The first and third terms are equal to zero because ψ_1 is a solution, and the same is true for the underlined terms because ψ_2 is a solution. Thus ψ is a solution. Finally, ψ can, of course, be normalized again by multiplying by a suitable constant.

In the case of the three p functions we can take any linear combination we like of $xf(r)$, $yf(r)$, $zf(r)$. We shall see in the next section that the choice

$$\begin{aligned}\psi_{+1} &= (x+iy)f(r), \\ \psi_{-1} &= (x-iy)f(r), \\ \psi_0 &= zf(r) \quad (i = \sqrt{(-1)})\end{aligned} \quad (14)$$

is a particularly important combination. The three functions (14) are connected with the angular momentum of the electron and with the Zeeman-effect.

Another example of a two-fold degeneracy is the two wave functions (sin and cos) for a free electron ((3), (4) of Chapter II). Both belong to the same energy $E = h^2/2m\lambda^2$.

THE HYDROGEN ATOM

In Chapter II (5) we have already formed the linear combination

$$\psi = \cos\frac{2\pi x}{\lambda} + i\sin\frac{2\pi x}{\lambda} = e^{2\pi ix/\lambda}.$$

This wave function also belongs to the same energy. The reason for this degeneracy is here that the electron can travel in the $+x$ and $-x$ directions (symmetry with respect to a reflection of the x-axis).

IV

ANGULAR MOMENTUM. ZEEMAN-EFFECT. SPIN

1. Sharp and unsharp quantities

In Chapter III we have studied different types of states (s, p, d states) whose wave functions, if expressed in polar coordinates, depend in a different way on the angles. The question arises as to what the physical significance of these different types of states is. We shall see that they are distinguished by the *angular momentum* which the electron has in these states. Before, however, such a question can be answered we must remember that in quantum mechanics a physical quantity does not always have a sharp value. For instance, as far as we have developed the theory at present, the energy has a sharp value, whereas the position of the electron is not sharp, there is only a probability distribution for it.

We now speak about another physical quantity, namely, the angular momentum of the electron, and we must ask whether or in what circumstances this quantity has a sharp value or whether only a probability distribution exists for the various values of this quantity. We decide this question by analogy with what we already know about the energy and the way in which the wave equation is connected with the classical expression for the energy. In Chapter II we have seen that

$$E = \frac{1}{2m}(p_x^2 + p_y^2 + p_z^2) + V(x, y, z) \qquad (1)$$

goes over into the wave equation if we replace p_x by

the operator

$$p_x \to \frac{\hbar}{i} \frac{\partial}{\partial x} \qquad (2)$$

and let the operator on the right side of (1) operate on a wave function $\psi(x,y,z)$. The coordinates in V are not replaced by any operators. Let us then, after this replacement, denote the right side of (1) by E_{operator}. The resulting wave equation $E\psi = E_{\text{operator}}\psi$ will have certain solutions ψ_n, belonging to the energy values E_n, say. In each case the energy has a *sharp value* E_n. E_{operator} acting on ψ_n produces here the *same* wave function ψ_n, only multiplied by the *constant* E_n. The result is $E_n\psi_n$.

It is now quite plausible to generalize this result in the following way: Suppose $Q(x,y,z,p_x,p_y,p_z)$ is a certain physical quantity which in general can be expressed by the coordinates and the momenta. If we now carry out the same replacement (2) of p_x, p_y, p_z by the operators $\frac{\hbar}{i}\frac{\partial}{\partial x},...,$ etc., Q becomes an operator Q_{operator}, say. Now let Q_{operator} operate on the wave function ψ. It may or may not then happen that Q_{operator} reproduces the wave function ψ with a certain factor q, i.e.

$$Q_{\text{operator}}\psi = q\psi, \qquad (3)$$

where q is a *constant* number (not an operator). In the case where (3) holds we are justified—by the analogy to the energy—in saying that the *physical quantity Q has a sharp value, namely, the value q*. On the other hand, we must expect in many cases that $Q_{\text{operator}}\psi$ will be a function of (x,y,z) quite different from ψ itself; in this case Q *does not have a sharp value* and there will only be a probability distribution for the various values of Q. This probability distribution can in fact be calculated, but this requires a further deepening of the theory which goes beyond the framework of this book. An example of the latter case is the coordinate x itself. $x\psi(x,y,z)$ is a function of the coordinates totally different from $\psi(x,y,z)$ itself and not just ψ multiplied by a *constant* factor. Therefore x does not have a sharp value.

As a further example we consider waves with *sharp momentum p_x*. p_x is expected to have a sharp value for a monochromatic wave of a free electron, because such a wave has a well-defined wave-length and there is a unique relationship between wave-length and velocity (or momentum). The wave functions for such a wave have been given in Chapter II, equations (3) and (4). However, if we apply the operator $p_x = \dfrac{\hbar}{i}\dfrac{\partial}{\partial x}$ to these wave functions, for instance to the sin solution

$$p_{x\ \text{operator}} \sin\frac{2\pi x}{\lambda} = \frac{\hbar}{i}\frac{\partial}{\partial x}\sin\frac{2\pi x}{\lambda} = \frac{2\pi\hbar}{i\lambda}\cos\frac{2\pi x}{\lambda},$$

the wave function is not reproduced, contrary to what we expect. The same is true for the cos solution. The reason is this: The sin and cos solutions are *standing waves* and describe an electron moving to and fro along the x-axis in both directions (for instance between two walls). Thus p_x occurs with both signs and cannot be expected to be sharp itself; only the absolute value or p_x^2 is sharp, and this is easily seen to be the case. But we can also easily construct a wave function for which p_x itself is sharp. We have only to remember that both solutions (sin and cos) are degenerate and belong to the same energy. We are therefore free to form linear combinations, for instance the combination Chapter II, eqn. (5):

$$\psi = \cos\frac{2\pi x}{\lambda} + i\sin\frac{2\pi x}{\lambda} = e^{2\pi i x/\lambda}.$$

ANGULAR MOMENTUM. ZEEMAN-EFFECT

This solution, which also satisfies the wave equation, has the required property

$$p_{x\ \text{operator}}\, e^{2\pi i x/\lambda} = \frac{2\pi \hbar}{\lambda} e^{2\pi i x/\lambda} = \frac{2\pi \hbar}{\lambda} \psi.$$

The numerical value of p_x is the coefficient of ψ on the right-hand side, $p_x = 2\pi\hbar/\lambda$, and this is just our old relation between wave-length and momentum. The linear combination

$$\cos\frac{2\pi x}{\lambda} - i\sin\frac{2\pi x}{\lambda} = e^{-2\pi i x/\lambda}$$

describes a particle travelling with momentum $2\pi\hbar/\lambda$ in the $-x$ direction.

A *wave packet* consisting of a superposition of several monochromatic waves with different wave-lengths cannot have a sharp momentum. Our criterion verifies this. For instance,

$$\psi = a e^{2\pi i x/\lambda_1} + b e^{2\pi i x/\lambda_2} \quad (\lambda_1 \neq \lambda_2), \tag{4}$$

$$\frac{\hbar}{i}\frac{\partial}{\partial x}\psi = \frac{2\pi\hbar}{\lambda_1} a e^{2\pi i x/\lambda_1} + \frac{2\pi\hbar}{\lambda_2} b e^{2\pi i x/\lambda_2}$$

(a and b may still depend on time). The right-hand side of the last equation is not proportional to ψ.† This is quite generally true: If a wave function ψ consists of several parts, say $\sum_k \psi_k$ such that for each part the quantity Q has a sharp value q_k but the q_k's are different, then Q cannot be sharp for a state described by the wave function ψ. The statement also holds the other way round: If a quantity Q

† The wave packet equation (4) does not even have a sharp energy $\left(\text{apply the operator } \dfrac{p_x^2}{2m} = -\dfrac{\hbar^2}{2m}\dfrac{\partial^2}{\partial x^2}\right)$. (4) is not a solution of the wave equation, (17), Chapter II, whose solutions always have a sharp energy. (4) is, however, a solution of the time-dependent wave equation, Chapter II, (22), provided a and b depend on time in a suitable way (compare, for example, Chapter II, (24)–(26)).

is not sharp in a state with wave function ψ, then ψ can be considered as composed of several parts ψ_k with different sharp values for each part. Examples are given in the following sections.

The reader will also easily find that the momentum p_x is not sharp for any of the hydrogen wave functions given in Chapter III, in agreement with the considerations on pp. 27, 36.

It is seen that the wave function of a particle with a given sharp value of the momentum is necessarily *complex*. The occurrence of complex wave functions is quite general in wave mechanics. This shows that ψ itself is not a physical quantity that could be measured directly in some way or other, as is the case, for instance, for an electric field. Its role is mathematical, but for that reason none the less vitally important for the description of nature. Only ψ^2 or $|\psi|^2$ (if ψ is complex) has a direct physical meaning, being the probability distribution for the position of the electron.

2. Angular momentum of s and p states

In classical mechanics the angular momentum **M** is a vector given by

$$\mathbf{M} = m(\mathbf{r} \times \mathbf{v}) = (\mathbf{r} \times \mathbf{p}),$$

where **v** is the velocity and **r** the radius from the origin of the coordinate system. $\mathbf{r} \times \mathbf{v}$ denotes the vector product. If the motion is in the xy-plane we may introduce polar coordinates

$$x = r\cos\phi, \qquad y = r\sin\phi,$$
$$\frac{dx}{dt} = v_x = -r\sin\phi\frac{d\phi}{dt} + \cos\phi\frac{dr}{dt},$$
$$\frac{dy}{dt} = v_y = +r\cos\phi\frac{d\phi}{dt} + \sin\phi\frac{dr}{dt}.$$

ANGULAR MOMENTUM. ZEEMAN-EFFECT

M has then only a z-component

$$M_z = m(xv_y - yv_x) = mr^2 \frac{d\phi}{dt}.$$

This is the well-known expression $mr^2 \times$ angular velocity. For a motion in three dimensions all three components M_x, M_y, M_z will be different from zero. They are:

$$M_x = yp_z - zp_y, \qquad M_y = zp_x - xp_z,$$
$$M_z = xp_y - yp_x. \tag{5}$$

In classical mechanics, if a particle moves in a field that is spherically symmetrical, all three components of **M** are *constants of the motion*, i.e. they remain constant in course of time, a fact that is well known, for instance from the motion of a planet round the sun. Is this also true in quantum mechanics? Clearly, before such a question can be raised, we must decide whether M_x, M_y, M_z have sharp values, for if this is not the case we cannot say that these quantities have constant values.†

In wave mechanics we again replace p_x, etc., by (2). The angular momenta then become operators:

$$M_x = \frac{\hbar}{i}\left(y\frac{\partial}{\partial z} - z\frac{\partial}{\partial y}\right), \qquad M_y = \frac{\hbar}{i}\left(z\frac{\partial}{\partial x} - x\frac{\partial}{\partial z}\right),$$
$$M_z = \frac{\hbar}{i}\left(x\frac{\partial}{\partial y} - y\frac{\partial}{\partial x}\right). \tag{6}$$

In order to find out whether the angular momenta have sharp values and, if so, what their values are, we let the operators (6) act on the wave function ψ. In particular

† Even if M_x is not sharp there is a certain meaning in saying that it is a constant of the motion. What then remains unchanged in course of time is the probability distribution for the various values of M_x. In this sense all three components of **M** are constants of the motion, also in wave mechanics.

52 ANGULAR MOMENTUM. ZEEMAN-EFFECT

we are interested in the values of the M's which the electron has in an s, p, or d state. The wave functions of an s state are functions of r only, for those of a p state we choose the wave functions, Chapter III (14), ψ_{+1}, ψ_0, ψ_{-1}.

Let the operators (6) then act on ψ and let us see whether, for instance, $M_z \psi = \lambda \psi$, where λ is the multiplying constant.

Let ψ first be an s wave function, i.e. $\psi = \psi(r)$,

$$\frac{i}{\hbar} M_z \psi = x \frac{\partial}{\partial y} \psi - y \frac{\partial}{\partial x} \psi$$

$$= x \frac{\partial \psi}{\partial r} \frac{y}{r} - y \frac{\partial \psi}{\partial r} \frac{x}{r} = 0.$$

In a similar way $M_x \psi = M_y \psi = 0$. Thus $\lambda = 0$ for an s state. We have therefore found that in an s state, *all three components of the angular momentum are sharp and have the value zero.*

Let now ψ be one of the three p wave functions, Chapter III (14). We shall show that we obtain three different values for M_z for the three p functions

$$\psi_{+1} = (x+iy)f(r),$$
$$\psi_{-1} = (x-iy)f(r),$$
$$\psi_0 = zf(r).$$

Differentiating ψ_{+1} we get

$$M_z \psi_{+1} = \frac{\hbar}{i} x \left(if + (x+iy) \frac{y}{r} \frac{\partial f}{\partial r} \right) - \frac{\hbar}{i} y \left(f + (x+iy) \frac{x}{r} \frac{\partial f}{\partial r} \right)$$

$$= \hbar \left(x - \frac{1}{i} y \right) f = \hbar (x+iy) f = \hbar \psi_{+1}.$$

ANGULAR MOMENTUM. ZEEMAN-EFFECT

This is an equation of the form $M_z\psi = \lambda\psi$, the multiplying constant λ being $+\hbar$. Therefore in the state ψ_{+1}, M_z has the sharp value $+\hbar$.

As regards the other two wave functions, ψ_{-1} and ψ_0, we easily find in the same way that

$$M_z\psi_{-1} = -\hbar\psi_{-1},$$

$$M_z\psi_0 = 0.$$

Therefore, in a p state the z-component of the angular momentum has the three values $+\hbar$, 0, $-\hbar$, if the electron has the wave functions ψ_{+1}, ψ_0, ψ_{-1} respectively.

If we now try to find the corresponding values of M_x and M_y, we shall find that $M_x\psi$ is *not* equal to $\lambda\psi$, for any of the three p functions ψ_{+1}, ψ_{-1}, ψ_0.

As an example, take

$$M_x\psi_0 = \frac{\hbar}{i}\left(y\frac{\partial}{\partial z} - z\frac{\partial}{\partial y}\right)zf(r)$$

$$= \frac{\hbar}{i}\left(yf + \frac{z^2y}{r}\frac{\partial f}{\partial r} - \frac{z^2y}{r}\frac{\partial f}{\partial r}\right) = \frac{\hbar}{i}yf(r).$$

$yf(r)$ is not the same function of x, y, z as $\psi_0 = zf(r)$. Therefore M_x *does not have a sharp value* if the electron is in the state ψ_0. The same is true for M_y and for the other two p functions ψ_{+1}, ψ_{-1}.†

---

† If we had chosen the wave functions ψ_x and ψ_y instead of ψ_{+1} and ψ_{-1} then M_z would not be sharp. ψ_x is a linear combination of ψ_{+1} and ψ_{-1}:

$$\psi_x = \tfrac{1}{2}(\psi_{+1} + \psi_{-1}).$$

This illustrates the statement made in section 1, namely, that if a quantity does not have a sharp value for a wave function ψ, then ψ consists of parts with sharp values for each part. In this case ψ_x consists of two parts with the values $+\hbar$ and $-\hbar$ for M.

54   ANGULAR MOMENTUM. ZEEMAN-EFFECT

We summarize our results in Table 1.

TABLE 1. *Angular momentum in s and p states*

State	M_z	M_x	M_y
s	0	0	0
$p\begin{cases}\psi_{+1}\\\psi_0\\\psi_{-1}\end{cases}$	$+\hbar$ 0 $-\hbar$	\}not sharp	

One may wonder why M_z plays a role so different from M_x and M_y since, after all, in the wave equation no particular direction is distinguished from any other direction. This is due to our particular choice of the linear combinations of the three degenerate p wave functions. We could equally well choose three different linear combinations, for instance $(y+iz)f(r)$, $(y-iz)f(r)$, $xf(r)$ changing $x \to y$, $y \to z$, $z \to x$. For these three wave functions M_x has sharp values, and they are also $+\hbar$, $-\hbar$, 0, whilst M_z and M_y are now unsharp. Thus we find in general: *In a p state only one component of the angular momentum can have a sharp value, and then has one of the three values $+\hbar$, 0, $-\hbar$, the other two components are then not sharp. The particular direction in which the angular momentum is sharp can be chosen arbitrarily.*

We also see that, as far as the angular momentum has a sharp value at all, it is *constant*. For $\hbar$ and zero are surely constants that do not change in course of time. Later we shall use the values of M_z to classify and label the quantum states.

3. Zeeman-effect

The arbitrariness as to the choice of the axis around which the angular momentum is sharp can be removed if we consider the electron moving in a weak *magnetic field*,

ANGULAR MOMENTUM. ZEEMAN-EFFECT

H, say. To understand what happens in this case consider first the classical case of an electron moving on a circle in the plane perpendicular to the direction of the field. Such an electron has then a *magnetic dipole*. The dipole moment is, as is well known,

$$\mu = \frac{e}{2c} rv,$$

if r is the radius of the circle, v the velocity of the electron, and c the velocity of light. Now the angular momentum **M** of the electron is $m(\mathbf{r} \times \mathbf{v})$, and if we call the direction of the magnetic field the z-axis, we can write $M_z = mrv$, or

$$\mu_z = \frac{e}{2mc} M_z. \tag{7}$$

This expression holds quite generally, even if the electron does not move on a circle or in a plane which is perpendicular to the magnetic field, M_z being then, of course, the component of the angular momentum in the z direction. It is very plausible—and can in fact be proved—that (7) should also hold in quantum theory, if only for M_z the value is substituted which the electron has according to quantum theory. In the magnetic field H the electron has an additional energy

$$W = \mu H = \frac{e}{2mc} M_z H. \tag{8}$$

This energy is, of course, in addition to the usual kinetic and potential energies occurring in the wave equation. Now the energy, being an *observable* quantity, has a sharp value. It is therefore obvious that we must choose the direction in which the angular momentum is sharp to be the direction of the field H. The measurement of the energy in the magnetic field can be regarded as a measurement of M_z forcing M_z to assume a sharp value.

If an atom in a p state is placed in such a field, it will have different energies according to the three different values of M_z, namely,

$$W = +\frac{e}{2mc}\hbar H \atop \text{or} \quad 0 \atop \text{or} \quad -\frac{e}{2mc}\hbar H \right\} . \qquad (9)$$

If we plot the energy levels in a magnetic field we see that the p level splits up into three levels (Fig. 13) with

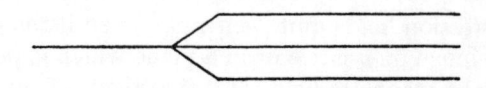

FIG. 13. Zeeman-effect of p state.

a separation given by formula (9). This splitting up of a single level into three in a magnetic field is called the 'Zeeman-effect', after Zeeman, who was the first to discover that certain spectral lines emitted by atoms in a strong magnetic field split up into several lines. The separation is proportional to the field strength H. Formula (8) has been fully justified by the experiments, at any rate for those levels which split up into three levels (compare, however, the following sections).

4. d states, directional quantization

In addition to s and p states there are also states with higher angular momentum. In Chapter III, section 3, for instance, d states have been mentioned. In a d state the

ANGULAR MOMENTUM. ZEEMAN-EFFECT

z component of the angular momentum M_z can have the five sharp values (as we shall not show in detail)

$$M_z = 2\hbar, \quad \hbar, \quad 0, \quad -\hbar, \quad -2\hbar.$$

(See Table 2.) In a magnetic field a d level would accordingly split up into five levels. There are also levels with still higher angular momentum splitting up into 7, 9,..., etc., levels, but these are of lesser importance, at any rate for light atoms.

It is often convenient to represent the angular momentum by the picture of *directional quantization*. We define an absolute value $\hbar l$ of the angular momentum as being the maximum value which M_z can assume; i.e.

$$\begin{array}{cccc} & s & p & d \text{ states} \\ l = & 0 & 1 & 2 \end{array}$$

and represent for each state the angular momentum by a vector of length l. Relative to the z direction l can have only a certain *discrete set of directions*, namely, such directions that the component of l in the z direction assumes the values $M_z/\hbar$.

TABLE 2. *Angular momentum of d state*

	M_z	M_x	M_y	l
d	$+2\hbar$ $+\hbar$ 0 $-\hbar$ $-2\hbar$	not sharp	not sharp	2

The 'directional quantization' picture is shown in Fig. 14. The various components of the orbital angular momentum are $\hbar$, 0, $-\hbar$ for a p state, and $2\hbar$, $\hbar$, 0, $-\hbar$, and $-2\hbar$ for a d state respectively. In all there are $2l+1$

different directions for l. This is also the number of levels into which the original level splits in the Zeeman-effect in a magnetic field. The values of M_z always differ by an integral multiple of $\hbar$ and are: $\hbar l, \hbar(l-1), ..., -\hbar(l-1), -\hbar l$.

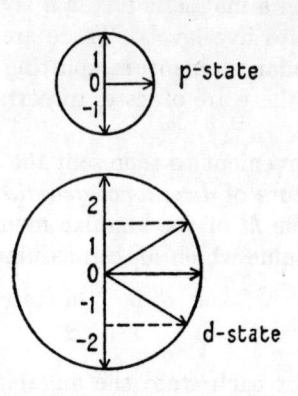

FIG. 14. Directional quantization.

The number l, defined as the maximum value of $M_z/\hbar$, has also a slightly different significance. We ask what the absolute value of the angular momentum M is, or its square

$$\mathbf{M}^2 = M_x^2 + M_y^2 + M_z^2.$$

The reader may easily verify that $\mathbf{M}^2$ has a sharp value for s and p states, namely 0 and $2\hbar^2$ respectively. This is generally so for an electron in a central field of force. The general formula is

$$\mathbf{M}^2 = \hbar^2 l(l+1). \tag{10}$$

We might have expected to find $\hbar^2 l^2$. l was the maximum value of $M_z/\hbar$ and when M_z is a maximum we would imagine the other components to be zero. This is not so because the other two components of $\mathbf{M}$ are not sharp and do not vanish. $M_x^2 + M_y^2$ has the sharp value $\mathbf{M}^2 - M_z^2 = \hbar^2 l$. The formula (10) has been confirmed by all experiments.

5. The electron spin

The question now arises whether all the above results are in perfect agreement with the experiments. The answer is only partly yes, and this will lead us to a further expansion of the theory.

The Schrödinger equation always gives perfect results if there is no magnetic field present. In the presence of a magnetic field it is true that cases exist where a level splits up into three or five levels. There are, however, also cases observed where a level splits up into only *two levels*. This is indeed the case for all atoms with only one valency electron, lithium, sodium, etc. A splitting up into two or four levels can never be obtained from the wave equation, which always leads to states with a 1, 3, 5,...-fold degeneracy.

The problem was solved by Goudsmit and Uhlenbeck (1925) by an ingenious suggestion. These authors assumed that the electron, in addition to its orbital motion, is capable of carrying out itself a *spinning motion*. For the moment we may picture the electron as a small solid body, rather than a point, carrying out a rotation about its own axis similar to that of a gyroscope. Thus in addition to its orbital angular momentum l the electron has its own angular momentum which we call the *spin* and which we denote by s. The value of s must be chosen so that we can account for the splitting into two levels in a magnetic field.

The orbital angular momentum l gives rise to $2l+1$ levels in a magnetic field. The spin angular momentum will therefore give rise to $2s+1$ levels. In order to make this equal to two, $2s+1 = 2$, s must have the value $\frac{1}{2}$. The component of the spin momentum in the z direction can then have the two values only:

$$M_{sz} = \begin{matrix} +\tfrac{1}{2}\hbar \\ -\tfrac{1}{2}\hbar, \end{matrix}$$

the subscript s refers to the 'spin momentum'.

In the picture of directional quantization (Fig. 15) there are only two directions for the spin; namely, the direction of the field itself or the opposite direction. In a magnetic field the spin alone will split up into two levels with equal separation from the original level.

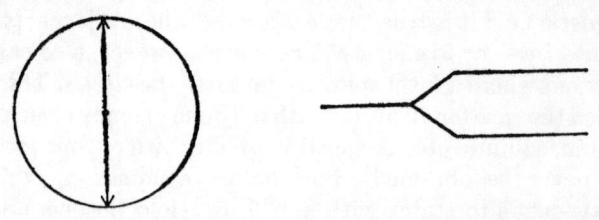

FIG. 15. Spin: directional quantization and Zeeman-effect.

From the very fact that the spin causes a splitting in a magnetic field it follows that a *magnetic moment* is also attached to the spin, as must be expected from the picture of a charged rotating ball. From the magnitude of the splitting the value of the magnetic moment can be obtained. It is found to be the same as that of an electron with angular momentum $M_z = \hbar$, namely (7),

$$\mu_s = \frac{e\hbar}{2mc}. \tag{11}$$

One might have expected that it is only half that value because the mechanical momentum of the spin is only $\hbar/2$. But this is not the case. The explanation of this factor cannot be given here. The spin and all its properties are now understood as a necessary and natural consequence of the theory of relativity combined with wave mechanics.

ANGULAR MOMENTUM. ZEEMAN-EFFECT

There the spin appears no longer as an independent addition to wave mechanics but is derived from the principles of these two theories. The value (11) of the magnetic moment follows then also. However, the picture of a rotating ball is not quite substantiated in this theory but the proper explanation goes beyond the framework of this book.

Since all angular momenta occurring in atomic physics are simple numbers multiplied by $\hbar$, it is convenient to introduce these numbers themselves instead of M_z. We denote them by m_z, etc:

$$M_z/\hbar = m_z, \qquad M_{sz}/\hbar = m_{sz}, \quad \text{etc.}$$

For simplicity we shall henceforward use the m_z, etc., and simply call them 'angular momentum'. Such numbers as l, m_z, m_{sz}, etc., which can be used to label the quantum states, are called *quantum numbers*.

If the electron is in an s state the orbital momentum l is zero and the spin is the only angular momentum of the electron. In this case the level actually splits up into two levels only. If, however, the electron has an orbital angular momentum $l > 0$, the result becomes more complicated, then the orbital momentum and spin momentum combine. In order to see in what way this combination takes place we consider the z components of the two momenta and denote the orbital momentum now by m_{lz}. The total angular momentum is obviously the sum of the two

$$m_z = m_{lz} + m_{sz}.$$

For a p state $l = 1$, m_z can have one of the following six values:

$$m_z = \begin{array}{|c|} 1 \\ 0 \\ -1 \end{array} + \begin{array}{|c|} +\tfrac{1}{2} \\ -\tfrac{1}{2} \end{array} = \tfrac{3}{2},\ \tfrac{1}{2},\ \tfrac{1}{2},\ -\tfrac{1}{2},\ -\tfrac{1}{2},\ -\tfrac{3}{2}.$$

The values $\frac{1}{2}$ and $-\frac{1}{2}$ occur twice, as they can be obtained in two ways: $\frac{1}{2} = 1-\frac{1}{2}$ or $\frac{1}{2} = 0+\frac{1}{2}$.

Now m_z is the z component of the *total* angular momentum, and it must therefore be possible to consider it as the component of a new vector denoting the *absolute value of the total angular momentum* corresponding to the l of the orbital momentum and the s of the spin. This quantity we denote by j. It represents the total angular momentum arising from the combination of spin and orbital momentum. The components of j, i.e. m_z, must be similar to m_{lz},

$$j, j-1, ..., -j.$$

Now, with the above values of m_z, there must be two different values for j, namely, $j = \frac{3}{2}$ with $m_z = \frac{3}{2}, \frac{1}{2}, -\frac{1}{2}, -\frac{3}{2}$ and $j = \frac{1}{2}$ with $m_z = \frac{1}{2}, -\frac{1}{2}$ (Table 3).

TABLE 3. *Combination of spin and orbital momentum*

m_z	$\frac{3}{2}$, $\frac{1}{2}$, $-\frac{1}{2}$, $-\frac{3}{2}$	$\frac{1}{2}$, $-\frac{1}{2}$
j	$\frac{3}{2}$	$\frac{1}{2}$

Thus an electron in a p state can actually exist in two different states with total angular momentum $\frac{3}{2}$ or $\frac{1}{2}$. The first is four-fold degenerate, the second two-fold; the degree of degeneracy is $2j+1$ in each case. This can be represented in the directional quantization picture, Fig. 16. The electron can put its *spin either in the direction of its orbital momentum or in the opposite direction*.

Actually, the two states have slightly different energies even in the absence of a magnetic field, for the following reason: The electron rotating in its orbit is equivalent to an electric circular current and therefore produces a weak

ANGULAR MOMENTUM. ZEEMAN-EFFECT

magnetic field itself. This interacts with the spin momentum forcing it, as in Fig. 16, either in the direction of this magnetic field (which is the same as the direction of the orbital momentum) or in the opposite direction. The energy in the two cases is, of course, slightly different.

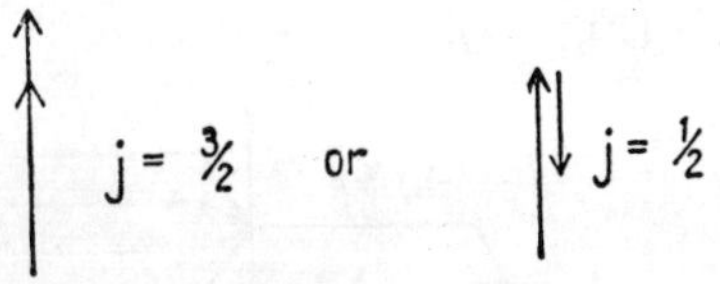

FIG. 16. Vector addition of spin and orbital momentum.

We estimate the order of magnitude of this splitting compared with the ordinary distances of the energy levels in an atom. The magnetic action of the orbit is that of a magnetic dipole placed at its centre with a moment (equation (7)) of the order of magnitude $e\hbar/mc$. The spin also has a magnetic moment of the same magnitude, but is, of course, placed on the orbit. Two magnetic dipoles at a distance r have a mutual potential energy of the order μ^2/r^3. For r we may insert a distance of the order of magnitude of the Bohr radius $1/a = \hbar^2/me^2$. Thus the energy-splitting is of the order of magnitude

$$\Delta E = \left(\frac{e\hbar}{mc}\right)^2 a^3 = e^2 a \left(\frac{e^2}{\hbar c}\right)^2. \qquad (12)$$

Here we have left one factor $e^2 a$. For this is the potential energy of the electron in its orbit and is the order of magnitude of the atomic energies. In (12), then, the factor $e^2/\hbar c$ occurs. This is a pure number without physical dimensions. (e^2 has the dimension energy × length, $\hbar$ that of energy × time, and c that of length/time.) Its numerical value, calculated from the known values of e, $\hbar$, c, is

1/137.† Thus, from (12) it follows that the splitting due to the spin-orbit coupling is 137^2 times smaller than the ordinary energies of an atom, which is very small. This splitting is called the *fine structure* of the spectral lines, and $e^2/\hbar c$ the *fine structure constant*.

If also an external field is applied each level j splits up into $2j+1$ levels (Fig. 17).

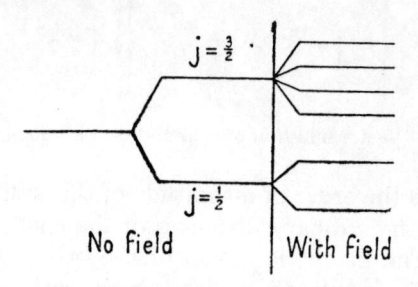

FIG. 17. Splitting of p level with spin.

An electron in a d state will have a total angular momentum $j = 2+\frac{1}{2} = \frac{5}{2}$ or $j = 2-\frac{1}{2} = \frac{3}{2}$. The first splits in a magnetic field into six, the second into four levels

6. Two electrons with spin

So far we have considered systems with one electron only, but of course most atoms and molecules have several electrons. We consider now the question of how the spins and orbital angular momenta of two electrons are to be combined. In a spherically symmetrical field (atom) each electron has spin $s_1 = s_2 = \frac{1}{2}$ and an orbital momentum l_1 and l_2 respectively.

† A very strange number indeed, remembering that it is built up of universal constants only. Its explanation is one of the greatest unsolved problems of modern physics.

ANGULAR MOMENTUM. ZEEMAN-EFFECT

Consider first the simple case in which there is no angular momentum, i.e. where l_1 and l_2 are both zero. Both electrons are then in s states. The two spins then combine to give a resultant spin for the total system. The way in which they combine is exactly the same as in the case of a spin-orbital momentum combination (section 5). The two spin components in the z direction are

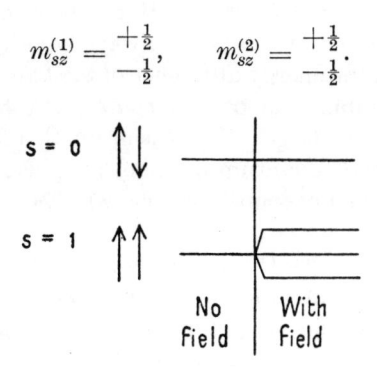

$$m_{sz}^{(1)} = \genfrac{}{}{0pt}{}{+\tfrac{1}{2}}{-\tfrac{1}{2}}, \qquad m_{sz}^{(2)} = \genfrac{}{}{0pt}{}{+\tfrac{1}{2}}{-\tfrac{1}{2}}.$$

FIG. 18. Zeeman-effect of singlets and triplets.

The resultant m_z is the sum of these, giving the four possible values
$$m_z = 1, \quad 0, \quad 0, \quad -1.$$

In the same way as in section 5, we conclude that the combination of the two spins gives rise to two different states with a total spin $s = 1$ and $s = 0$. In other words, the spins can have the same direction (↑↑) giving $s = 1$ or opposite directions (↑↓) giving $s = 0$.

In a magnetic field the state $s = 0$ will show no splitting, but the state $s = 1$ will split up into three levels much in the same way as a p state. The generally accepted notation for states with various total spins (irrespective of how many electrons the system has) is

$$s = 0, \quad \tfrac{1}{2}, \quad 1,$$
$$\text{singlets}, \quad \text{doublets}, \quad \text{triplets}.$$

One would think that the energy difference of the two states (singlet and triplet) arising from the combination is very small and due only to the small magnetic influence of one spin upon the other. But this is not the case. States with different total spin always have very different energies, and the energy difference of the two states arising from the combination of two spins (in a He atom, for instance) is very large. The reason for this fact is due to a fundamental quantum-mechanical phenomenon (the 'exchange phenomenon') which will be explained in Chapter V.

We now consider the more general case where one or both electrons (in different orbits) have also an orbital angular momentum l_1, l_2 respectively. The question arises then as to the *order* in which we have to combine all these momenta. As was mentioned above, states with different total spin have very different energies, in other words, the *coupling between the spins is always large*, whereas the coupling between spin and orbital momentum is small, giving rise to only a very small splitting. The coupling between the two orbital momenta of the two electrons is also usually fairly large. We therefore arrive at the following rule: *First combine the electron spins and obtain the total spins and also combine the two orbital momenta and obtain the total orbital momentum. Afterwards combine the total spin with the total orbital momentum to obtain the total angular momentum.* The energy splitting arising from the latter combination is usually very small (of the order of magnitude (12)).

The following example will make the rule clear. Let

ANGULAR MOMENTUM. ZEEMAN-EFFECT

$l_1 = l_2 = 1$ and $s_1 = s_2 = \frac{1}{2}$. The combination of the spins gives
$$s = s_1 \pm s_2 = 1, 0 \quad \text{(singlet and triplet).}$$

The two orbital momenta combined give a total orbital momentum which the reader will now easily find to be
$$l = 2, \quad 1, \quad 0.$$

To denote states with various total orbital momenta we use capital letters: S ($l = 0$), P ($l = 1$), D ($l = 2$), etc., while small letters like s, p, d refer to one single electron. We speak, for instance, of a 'singlet P state' written as ^1P if $l = 1$ and $s = 0$, and of a 'triplet S state' written as ^3S if $l = 0$ and $s = 1$, etc. For our example we find, therefore, that the following states arise out of a combination of $l_1 = l_2 = 1$, $s_1 = s_2 = \frac{1}{2}$:

$$^1S, \quad ^3S, \quad ^1P, \quad ^3P, \quad ^1D, \quad ^3D.$$

All these six states have rather different energies. Finally we may combine the total spin with the total orbital momentum to obtain the total angular momentum j. For the states ^1S, ^3S, ^1P, ^1D the combination gives nothing new because one of the two total momenta is zero. The total angular momenta for these states are therefore $j = 0, 1, 1, 2$ respectively. For ^3P ($l = 1$, $s = 1$) we find three states $j = 2, 1, 0$. For ^3D ($l = 2$, $s = 1$) we find also three states $j = 3, 2, 1$. The energy differences of these states with different j are usually very small and will play no role in the following. In a magnetic field each state with total momentum j will further split into $2j+1$ levels.

The rule for the combination of angular momenta explained above is called the *vector addition rule* (see also chapter VII, section 2.)

V

PROBLEM OF TWO ELECTRONS

1. The wave equation for two electrons

For the motion of an electron in a potential V we have found the wave equation

$$\nabla^2 \psi + \frac{2m}{\hbar^2}(E-V)\psi = 0,$$

$\psi = \psi(x, y, z)$. This equation, with ψ depending on the three coordinates of the electron, refers only to one single electron. In order to treat more complicated atoms or molecules we have to generalize it to describe two or more electrons. To obtain guidance as to how this can be done we consider again the classical expression for the energy,

$$E = \frac{p_x^2 + p_y^2 + p_z^2}{2m} + V(x, y, z).$$

If we now consider two electrons moving in the *same potential field* V the energy will be

$$E = \frac{p_{x_1}^2 + p_{y_1}^2 + p_{z_1}^2}{2m} + \frac{p_{x_2}^2 + p_{y_2}^2 + p_{z_2}^2}{2m} +$$
$$+ V(x_1, y_1, z_1) + V(x_2, y_2, z_2) + V_{12}, \quad (1)$$

where p_{x_1}, p_{x_2}, etc., are the momenta of the first and second electrons respectively, and x_1, x_2, y_1, y_2, etc., their coordinates. The potential V is the same function for both electrons, but the two electrons may, of course, have different positions. $V(x_1, y_1, z_1)$ is the potential energy of the first and $V(x_2, y_2, z_2)$ of the second electron. In addition the two electrons repel each other, which is accounted for by the mutual interaction potential

$$V_{12} = \frac{e^2}{r_{12}}, \qquad (2)$$

where r_{12} is the distance between the two electrons.

In order to translate equation (1) into wave mechanics we use the old method of replacing p_x by the operator $\frac{\hbar}{i}\frac{\partial}{\partial x}$ acting on a wave function ψ. To distinguish between the two electrons we have to replace p_{x_1} by $\frac{\hbar}{i}\frac{\partial}{\partial x_1}$ and p_{x_2} by $\frac{\hbar}{i}\frac{\partial}{\partial x_2}$ acting on ψ. Obviously, now, ψ must be a function of the coordinates of both electrons

$$\psi = \psi(x_1, y_1, z_1, x_2, y_2, z_2).$$

Hence

$$\frac{2m}{\hbar^2}E\psi = -\left(\frac{\partial^2}{\partial x_1^2}+\frac{\partial^2}{\partial y_1^2}+\frac{\partial^2}{\partial z_1^2}\right)\psi - \left(\frac{\partial^2}{\partial x_2^2}+\frac{\partial^2}{\partial y_2^2}+\frac{\partial^2}{\partial z_2^2}\right)\psi + \frac{2m}{\hbar^2}V\psi, \qquad (3)$$

with
$$V = V(x_1, y_1, z_1) + V(x_2, y_2, z_2) + V_{12}$$
$$= V_1 + V_2 + V_{12}.$$

Using the abbreviations

$$\nabla_1^2 \quad \text{for} \quad \frac{\partial^2}{\partial x_1^2}+\frac{\partial^2}{\partial y_1^2}+\frac{\partial^2}{\partial z_1^2}$$

and
$$\nabla_2^2 \quad \text{for} \quad \frac{\partial^2}{\partial x_2^2}+\frac{\partial^2}{\partial y_2^2}+\frac{\partial^2}{\partial z_2^2},$$

we get
$$\nabla_1^2 \psi + \nabla_2^2 \psi + \frac{2m}{\hbar^2}(E-V)\psi = 0. \qquad (4)$$

This is the wave equation for two electrons. The potential function $V(x_1,...,x_2,...)$, (3), is a function of the coordinates of *both* particles. The wave function is also a function of all the coordinates of both particles $\psi(x_1, y_1, z_1, x_2, y_2, z_2)$, i.e. of six variables.

In analogy to the physical interpretation of ψ^2 for one electron (Chapter I, section 3) we interpret now $\psi^2(x_1, y_1, z_1, x_2, y_2, z_2)$ as the *probability for finding the first electron at the position x_1, y_1, z_1 and at the same time the second electron at the position x_2, y_2, z_2*.

2. Solution of the wave equation for two electrons

We have seen that for two electrons the Schrödinger equation becomes

$$\nabla_1^2 \psi + \nabla_2^2 \psi + \frac{2m}{\hbar^2}\left(E - V_1 - V_2 - \frac{e^2}{r_{12}}\right)\psi = 0. \tag{5}$$

The exact solution of such an equation is always very difficult. In order to solve it we must make approximations. As a first approximation we neglect the term e^2/r_{12}. In other words, we regard the electrons as *moving independently of each other* in an external potential field. We should then expect that each electron can be in certain energy-levels irrespective of the energy-level occupied by the other electron. These energy-levels are the same as if only one electron were present moving in the field V, namely, $E_1, E_2, ..., E_k, ...$, say. The two electrons can then occupy two of these levels, E_a and E_b, for instance, and the total energy will be $E_a + E_b$. Each electron can occupy any of the levels E_k, all distributions are possible. This will be exactly what we shall find by solving equation (5). We find a solution by putting ψ equal to the product of the two functions, e.g.

$$\psi = \psi_a(1)\psi_b(2), \tag{6}$$

where $\psi_a(1) \equiv \psi_a(x_1, y_1, z_1)$ depends on the coordinates of the first electron only, the index a being used to distinguish

PROBLEM OF TWO ELECTRONS 71

one particular kind of solution. $\psi_b(x_2, y_2, z_2)$ depends on the second electron only. We insert (6) into the wave equation (5). ∇_1^2 acts on $\psi_a(1)$ only, while $\psi_b(2)$ is a constant with respect to the differentiation ∇_1^2. The first term of (5) is then $\psi_b(2)\nabla_1^2\psi_a(1)$. On the other hand, ∇_2^2 acts on $\psi_b(2)$ only and the second term becomes $\psi_a(1)\nabla_2^2\psi_b(2)$. Substituting in (5) (always neglecting V_{12}) we get then

$$\psi_b(2)\nabla_1^2\psi_a(1)+\psi_a(1)\nabla_2^2\psi_b(2)+\frac{2m}{\hbar^2}E\psi_a(1)\psi_b(2)-$$

$$-\frac{2m}{\hbar^2}\psi_b(2)V_1\psi_a(1)-\frac{2m}{\hbar^2}\psi_a(1)V_2\psi_b(2)=0. \quad (6')$$

Consider now the case where only the electron 1 is present, moving, of course, in the field V_1. Its wave function will then satisfy the equation for one electron

$$\nabla_1^2\psi(1)+\frac{2m}{\hbar^2}(E-V_1)\psi(1)=0. \quad (7)$$

Since the two electrons are moving independently, and from what we have said above, it is clear that the $\psi_a(1)$ which occurs in (6) should be identical with some solution of (7). Thus $\psi_a(1)$ must satisfy the equation

$$\nabla_1^2\psi_a(1)+\frac{2m}{\hbar^2}(E_a-V_1)\psi_a(1)=0, \quad (8)$$

and we shall see that (6') is then really satisfied. Similarly, the second electron, if it were alone present, would satisfy exactly the same equation (7), only with the coordinates 1 and 2 interchanged:

$$\nabla_2^2\psi(2)+\frac{2m}{\hbar^2}(E-V_2)\psi(2)=0. \quad (9)$$

A particular solution is $E=E_b$, $\psi=\psi_b(2)$.

The complete set of solutions of (9) is identical with the complete set of solutions of (7) since it can make no difference whether we call a single electron No. 1 or No. 2. But for a certain special solution each electron may be in a different energy state E_a or E_b. In the same way as above we assume that $\psi_b(2)$ occurring in (6) is identical with some solution of (9):

$$\nabla_2^2 \psi_b(2) + \frac{2m}{\hbar^2}(E_b - V_2)\psi_b(2) = 0. \tag{10}$$

Making use now of the fact that ψ_a and ψ_b satisfy the equations (8) and (10) separately, we easily see that (6') is satisfied provided that E is suitably chosen. The first and fourth terms of (6') are together equal to

$$-\frac{2m}{\hbar^2} E_a \psi_a(1)\psi_b(2)$$

and the second and fifth together equal

$$-\frac{2m}{\hbar^2} E_b \psi_b(2)\psi_a(1).$$

Hence (6') reduces, on cancelling the common factor $\frac{2m}{\hbar^2}\psi_b(2)\psi_a(1)$, to

$$E = E_a + E_b. \tag{11}$$

Thus the *product of two wave functions* for one electron each

$$\psi = \psi_a(1)\psi_b(2)$$

is a solution of the wave equation for two electrons with no interaction between them, the *energy of the whole system is just the sum of the energies of the two electrons*, as was to be expected.

Let us consider, for instance, the helium atom. It has a nucleus with charge $2e$ and two electrons, 1 and 2. Each electron moves in a potential $V = -2e^2/r$, and can exist in a set of energy-levels for one electron (He$^+$ ion). This

PROBLEM OF TWO ELECTRONS 73

set of energy-levels is very similar to that of the H atom, the only difference being that the nuclear charge is 2e instead of e. (It can easily be seen from Chapter III that the energy of each level is then just four times the energy of the corresponding level of hydrogen.) A state of the He atom is then characterized by a distribution of the two electrons on the set of energy-levels of the

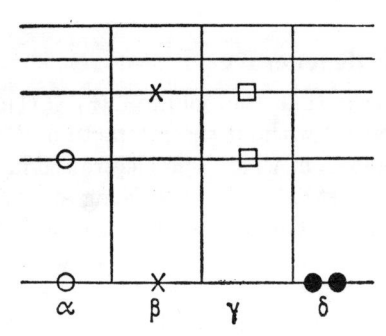

FIG. 19. Distribution of two electrons on one-electron levels.

He+ ion. Various such distributions are shown in Fig. 19. In distribution α one electron is in the ground state and the other in the first excited state. In γ both electrons are in excited states, and in δ both electrons are in the ground state. The lowest state of the atom is obviously the distribution δ.

For all these considerations we must not forget that we have neglected the mutual interaction of the two electrons. If the latter is taken into account the states of the atom can no longer be pictured in such a simple way as the distribution of electrons on the levels of the He+ ion.

The wave function ψ for the two electrons is no longer a function in three-dimensional space. It is a function of six coordinates $x_1, y_1, z_1, x_2, y_2, z_2$. If it were at all to be

3. Exchange degeneracy

We have seen that the solution (6) $\psi_a(1)\psi_b(2)$ of the wave equation (5) without the interaction e^2/r_{12} describes a state of the system where one electron is in the state a, the other in the state b, a and b being states of the one-electron problem. This is now not the only solution describing the same distribution of electrons. The reason is this: Actually, (6) describes the state where the electron labelled No. 1 is in level a and the electron No. 2 in b. Physically, however, the two electrons are *indistinguishable* from each other. This is also clear from the fact that the wave equation (5) is completely symmetrical in the two electrons 1 and 2.

We therefore expect that

$$\psi = \psi_a(2)\psi_b(1) \qquad (12)$$

is also a solution of (5), and this is easily verified. The solution (12) also belongs to the energy

$$E = E_a + E_b.$$

It describes the state where electron 2 is in a and electron 1 is in b. Hence there are two wave functions belonging to the same energy. We have to deal with a case of *degeneracy*, similar to that found before for a p state. The new kind of degeneracy is due to the possibility of

PROBLEM OF TWO ELECTRONS 75

exchanging the labels of the electrons, in other words, to the complete symmetry of the wave equation in the electron labels. We call it *exchange degeneracy*.

From (6) and (12) we can build up more wave functions belonging to the same energy. We may take any linear combination as was the case with the three p wave functions. For instance, we could choose the sum or difference of (6) and (12):

$$\psi_+ = \psi_a(1)\psi_b(2)+\psi_a(2)\psi_b(1), \qquad (13a)$$

$$\psi_- = \psi_a(1)\psi_b(2)-\psi_a(2)\psi_b(1). \qquad (13b)$$

(13) is also a solution of (5). (13 a) and (13 b) are particularly important combinations. These linear combinations have namely the property that ψ does not change essentially if we interchange 1 and 2. Obviously ψ_+ satisfies this requirement as it is symmetrical in 1 and 2. ψ_- changes its sign if 1 and 2 are interchanged, but $-\psi_-$ is not really a wave function different from ψ_-, since we can multiply every wave function by an arbitrary factor without changing its physical meaning. For both the solutions (13) there is therefore no real distinction between the two electrons.

The use of the linear combinations (13) instead of the mere products (6) and (12) is really compulsory. We must accept it as a very *fundamental principle that no distinction can be made between two electrons*. The probability of finding electron No. 1 at $x_1,...$ and electron No. 2 at $x_2,...$ must be equal to the probability of finding electron No. 1 at $x_2,..$ and No. 2 at $x_1,...$. All that has a physical meaning is the probability of finding *one* electron at $x_1,...$ and *one* electron at $x_2,...$ irrespective of any labels. Hence it follows that

$$\psi^2(1,2) = \psi^2(2,1),$$

and hence

$$\psi(1,2) = \psi(2,1) \quad \text{or} \quad \psi(1,2) = -\psi(2,1). \qquad (13')$$

The two wave functions (13) are just the ones which satisfy these two alternative conditions respectively.

The exchange degeneracy, of course, only exists so long as the interaction between the two electrons is neglected. Otherwise we shall see (Chapter VI) that (13a) and (13b) belong to different energies. If the interaction between the two electrons is taken into account the solution will be some wave function $\psi(1, 2)$ that can no longer be split up into products (6) or (12) or their combinations (13). On the other hand the symmetry properties (13') must be maintained. It is still true that no distinction can be made between the two particles and that therefore

$$\psi^2(1, 2) = \psi^2(2, 1),$$

no matter what $\psi(1, 2)$ is otherwise. So one of the symmetry properties (13') must hold for each solution but symmetrical and antisymmetrical solutions will in general have different energies.

4. The Pauli exclusion principle

So far we have been dealing with the orbital motion of the two electrons only and have not considered their spin. For a full description we must take into account the spin. As we have seen before, each electron has a spin momentum whose z component can have the two values $m_{sz} = +\frac{1}{2}$ or $-\frac{1}{2}$. We symbolize these two spin-states by arrows ↑ and ↓ respectively. The spins combine to a total spin $s = 0$ and 1, which we also symbolize by ↑↓ and ↑↑ respectively. States ↑↑ are three-fold degenerate.

Let us, then, consider the various states of two electrons, for instance the distributions α and δ of Fig. 19. To describe them fully we must draw an arrow through each electron showing the spin. It is obvious that we obtain then four times more cases, because each electron can have two spin directions. They are shown in Fig. 20. Cases ↑↑ are

PROBLEM OF TWO ELECTRONS 77

triplet states and ↑↓ are singlet states. (Each triplet comprises 3 degenerate states.) In cases where the two spins are parallel the atom has a magnetic moment due to the spin (cases α and γ). If the two spins are 'antiparallel' as in β or δ there is no magnetic moment. α and β

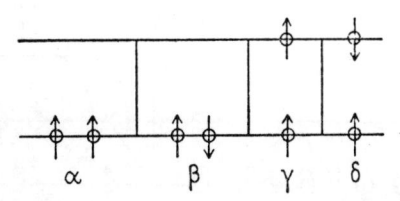

FIG. 20. He states, allowed and forbidden.

can be distinguished from each other experimentally, since α would split in a magnetic field into three states (see Chapter IV) whilst in β the spins cancel. The same is true for γ and δ. Without a magnetic field α and β have the same energies.

Whereas for atoms with only one electron it could be stated that, after the inclusion of the spin, the results of the theory are in perfect agreement with the experiments, one further fundamental principle has now to be added to the theory when we are dealing with two or more electrons.

We can compare our results with the experimental facts, for instance for the He atom. The ground state of the He atom could be either α or β; both have the same energy. It is now an experimental fact that the ground state of He does not split up in a magnetic field. Besides, He is known to be diamagnetic and never paramagnetic. This means that the He atom in its ground state has no magnetic moment. We must therefore conclude that the *state α (Fig. 20) does not occur in nature*. On the other hand, the spectroscopical evidence has shown that all the other states of Fig. 20, β, γ, δ, do occur in nature. It appears that we have to deal here with a new fundamental principle according to which not all the states derived from wave mechanics (generalized by the inclusion of the spin) are possible. The principle is due to Pauli (1925) and is called the *exclusion principle*.

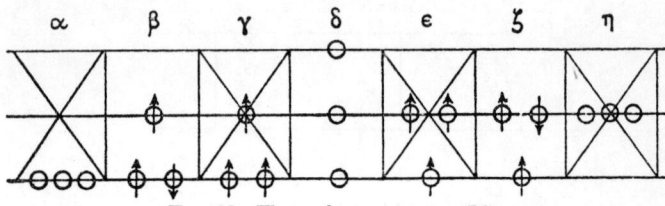

FIG. 21. Three-electron states (Li).

In order to find out which are the 'forbidden' and which the 'allowed' states, we consider the more general case of three electrons, for instance the Li atom. We can obviously describe the states of the atom again by distributing the electrons in all possible ways over the one-electron levels, i.e. the levels of the Li^{++} ion and drawing an arrow through each electron to describe its spin. Some of the distributions with low energy are given in Fig. 21. Distributions where no arrows are drawn stand for all the distributions with various spin directions. Now those distributions which are found not to occur in nature are crossed out. The exclusion of the states α, γ, ϵ, η was derived by Pauli from a careful study of the spectra of more complicated atoms. It is easy to read off from Fig. 21 the general rule: *All distributions with three electrons in the same level are forbidden. Two electrons may be in the same level, but must then have opposite spin directions.* For electrons in different levels there is no restriction of the spin directions.

We can also formulate this exclusion principle in a simpler way if we change what we call a 'state' of a single electron. We may include the description of the spin also in the description of its state and call the state of an electron different if either its spin or its orbit is different. The exclusion principle then assumes the simple form: *There can never be two (or more) electrons in the same state.* ✓

5. The spin wave function

The formulation of the Pauli principle given in the preceding section rests on the neglect of the interaction between the electrons. In general it is not possible to speak of a distribution of electrons amongst the one-electron levels and the wave function $\psi(1, 2)$ is not in general a product $\psi_a(1)\psi_b(2)$ or a linear combination of such products. We must look, therefore, for a more general formulation of the exclusion principle which can be applied also if the interaction between the electrons is taken into account. For this purpose it is useful to describe the spin also by a 'wave function', similar to the wave function $\psi_a(x, y, z)$ describing the orbital motion. Since there are two states of the spin of one electron, we shall have two wave functions α and β, say, corresponding to the values of the z component of the spin momentum m_{sz}, namely $+\frac{1}{2}$ and $-\frac{1}{2}$ respectively. The two spin functions α, β correspond to the three orbital wave functions of an electron in a p state.

If we picture the electron as a sphere rotating about the axis of the spin direction either from west to east or the other way round† the azimuthal angle ϕ of this rotation

---

† This picture of the spinning electron as a rotating ball must not be taken literally. No physical reality whatsoever can be attached to the 'structure of the electron'. The *only observable* fact is that the electron has an inner degree of freedom, the spin, and we use this picture merely to illustrate this fact. But no further conclusions should be derived from this picture and questions of what the 'radius' of such a ball would be, etc., are void of any physical meaning (compare p. 61).

may be the variable, upon which the spin functions α, β will depend. We need not know what functions of ϕ $\alpha(\phi)$ and $\beta(\phi)$ are. One may also use different variables. In all our following considerations the variable on which α, β depend will not occur; all we need is the fact that two such spin functions α, β exist.

The complete wave function of an electron is the product of the orbital wave function $\psi_a(x,y,z)$ multiplied by one of the spin functions:

$$\psi_a(x,y,z)\alpha, \qquad \psi_a(x,y,z)\beta. \qquad (14)$$

These describe an electron in the same orbital state a but with different spin directions. If we have two electrons, there will be two spin functions $\alpha(\phi_1)$, $\alpha(\phi_2)$, etc., or more shortly $\alpha(1)$ and $\alpha(2)$, etc. The total spin wave function is the product of two such spin functions (the spins move practically independently). In all we have, for two electrons, four spin functions:

	electron 1	electron 2
$\alpha(1)\alpha(2)$	↑	↑
$\beta(1)\beta(2)$	↓	↓
$\alpha(1)\beta(2)$	↑	↓
$\beta(1)\alpha(2)$	↓	↑

(15)

Again, we can form any linear combination of the four functions (15). To comply with the fact that the two electrons are indistinguishable from each other we shall again choose, just as in section 3, the symmetrical and antisymmetrical combinations of (15). $\alpha(1)\alpha(2)$ and $\beta(1)\beta(2)$ are already symmetrical. Instead of $\alpha(1)\beta(2)$ and $\alpha(2)\beta(1)$ we choose $\alpha(1)\beta(2)+\alpha(2)\beta(1)$ and $\alpha(1)\beta(2)-\alpha(2)\beta(1)$, of which the first is symmetrical, the second antisymmetrical. Now each α describes a spin with a z component $m_{sz} = +\frac{1}{2}$

PROBLEM OF TWO ELECTRONS

and β with $m_{sz} = -\tfrac{1}{2}$. The total spin in the z direction of the two electrons is then, for our four spin functions,

$$m_{sz} = m_{sz}^{(1)} + m_{sz}^{(2)}$$

$\alpha(1)\alpha(2)$	$+1$	$(16a)$
$\alpha(1)\beta(2) + \alpha(2)\beta(1)$	0	$(16b)$
$\beta(1)\beta(2)$	-1	$(16c)$
$\alpha(1)\beta(2) - \alpha(2)\beta(1)$	0	$(16d)$

These four spin functions are associated in a unique way with the total spin s of the two electrons. (See Chapter IV, section 6.) It will be remembered that two electrons with spins $\tfrac{1}{2}$ combine to give a resultant spin

$$s = 1 \text{ and } 0,$$

with the components

$$m_{sz} = 1, 0, -1, \text{ and } 0.$$

It is clear that the spin functions $(16a)$ and $(16c)$ belong to a total spin $s = 1$, because only $s = 1$ can have a component $m_{sz} = +1$ or -1. One of the two spin functions $(16b)$ or $(16c)$ belongs also to $s = 1$ and the other to $s = 0$. It seems plausible that $(16b)$ should belong to $s = 1$ and $(16d)$ to $s = 0$ because $(16b)$ is symmetrical and the other spin functions $(16a)$ and $(16c)$ which we already know to belong to $s = 1$ are also symmetrical. $(16d)$ is the only antisymmetrical spin function. The three spin functions of $s = 1$ belong to one and the same state (a triplet state), at any rate as long as there is no magnetic field present, and it is certainly plausible that they should all have the same symmetry properties.

That this must really be so follows from the following consideration: The three spin functions for $s = 1$ correspond closely to the three wave functions of a p state

82   PROBLEM OF TWO ELECTRONS

(Chapter III, section 3). They are degenerate and we are allowed to choose any linear combinations of them we like, whilst the spin functions for $s = 0$ and $s = 1$ belong to different energies. However, the spin functions, or any linear combinations of them, must still be perfectly either symmetrical or antisymmetrical in the two electrons. Suppose now the spin function (16 d) to belong to a total spin $s = 1$. We should then be entitled to form linear combinations of (16 d) and (16 a) since both would belong to $s = 1$. Such a combination would no longer be either symmetrical or antisymmetrical as it consists of two parts with different symmetry properties. Only combinations between (16 a, b, c) remain always symmetrical. It follows that (16 d) must stand apart and must be the wave function belonging to $s = 0$.

We thus arrive at the following scheme for the spin functions:

TABLE 4. *Symmetry of spin functions and total spin*

| Symmetrical | $s = 1$ ↑ ↑ | Triplet state (16 a–c) |
| Antisymmetrical | $s = 0$ ↑ ↓ | Singlet state (16 d) |

6. General formulation of the Pauli principle

It will now be easy to formulate the Pauli principle in a way that does not depend on the interaction between the electrons. We multiply the two parts of the wave functions for two electrons, the orbital wave functions (13), and the spin wave functions (16). We obtain altogether eight different wave functions:

$$[\psi_a(1)\psi_b(2)+\psi_a(2)\psi_b(1)]\begin{Bmatrix} \alpha(1)\alpha(2) \\ \beta(1)\beta(2) \\ \alpha(1)\beta(2)+\alpha(2)\beta(1) \end{Bmatrix} \quad (17a)$$

$$[\psi_a(1)\psi_b(2)+\psi_a(2)\psi_b(1)]\times[\alpha(1)\beta(2)-\alpha(2)\beta(1)] \quad (17b)$$

PROBLEM OF TWO ELECTRONS

$$[\psi_a(1)\psi_b(2)-\psi_a(2)\psi_b(1)]\begin{Bmatrix} \alpha(1)\alpha(2) \\ \beta(1)\beta(2) \\ \alpha(1)\beta(2)+\alpha(2)\beta(1) \end{Bmatrix} \quad (17c)$$

$$[\psi_a(1)\psi_b(2)-\psi_a(2)\psi_b(1)] \times [\alpha(1)\beta(2)-\alpha(2)\beta(1)]. \quad (17d)$$

We have written out these eight wave functions so that their symmetry properties are evident. $(17a)$ and $(17d)$ are symmetrical in the two electrons $((17d)$ changes its sign twice if 1 and 2 are commuted). $(17b)$ and $(17c)$ are antisymmetrical. Now according to the Pauli exclusion principle not all of these eight wave functions can be allowed.

To find out the forbidden wave functions we consider the case where the two electrons are in the same orbit, i.e. $a = b$. $(17c)$ and $(17d)$ vanish in this special case automatically. We know from the exclusion principle that in this case the two electrons must have opposite spin and the state is a singlet state. Therefore the three wave functions $(17a)$ are forbidden, and $(17b)$ is the one describing the state occurring. This wave function is *antisymmetrical*.

If the two electrons are in different orbits, one triplet state exists in nature in addition to the singlet state $(17b)$. It is very probable that this is not described by $(17a)$ (since for $a = b$ $(17a)$ is excluded) but by $(17c)$. $(17c)$ is also antisymmetrical in the two electrons. Furthermore, only *one* singlet state is known even if the two electrons are in different orbits. $(17b)$ has already been seen to be allowed, and we conclude therefore that $(17d)$ must also be forbidden. The allowed wave functions are therefore $(17b)$ and $(17c)$, and both are antisymmetrical. We are therefore led to the following principle: *The total wave function of two electrons is always antisymmetrical in the electrons*. If the wave function is split up into an orbital part and a spin part the orbital part can be symmetrical or antisymmetrical. The spin part is then antisymmetrical or symmetrical respectively.

The following table gives the allowed total wave functions for two electrons with no interaction either in two different orbits $a \neq b$ or in the same orbit $a = b$.

TABLE 5. *Symmetry of orbital wave functions and spin*

Orbital	Spin	Total s
$\psi_a(1)\psi_b(2)+\psi_a(2)\psi_b(1)$	$\alpha(1)\beta(2)-\alpha(2)\beta(1)$	0 (singlet)
$\psi_a(1)\psi_b(2)-\psi_a(2)\psi_b(1)$	$\left\{\begin{array}{c}\alpha(1)\alpha(2)\\ \beta(1)\beta(2)\\ \alpha(1)\beta(2)+\alpha(2)\beta(1)\end{array}\right\}$	1 (triplet)
$a=b \quad \psi_a(1)\psi_a(2)$	$\alpha(1)\beta(2)-\alpha(2)\beta(1)$	0 (singlet)

The new formulation of the exclusion principle is not confined to the case where the interaction of the electrons is neglected, nor is it confined to two electrons. For the symmetry of a wave function is a property that persists if the wave function is not merely a combination of products $\psi_a(1)\psi_b(2)$ but is some more general function $\psi(1, 2)$. Also for three or more electrons we can demand that the wave function be antisymmetrical in *all* electrons. To satisfy the exclusion principle we have to postulate that *every wave function of two or more electrons is, if the spin variables are included, antisymmetrical in all these electrons*, i.e. changes sign if the labels of any two electrons are exchanged. Thus for three or more electrons $\psi(1, 2, 3,...)$ $= -\psi(2, 1, 3,...) = -\psi(1, 3, 2,...)$, etc. If, for three electrons, the interaction is neglected, the wave function is built up of products such as $\psi_a(1)\psi_b(2)\psi_c(3)$, $\alpha(1)\alpha(2)\beta(3)$ and the antisymmetrical combination obtained by permuting the variables 1, 2, 3 is to be taken. We leave it to the reader to show that then the wave function vanishes automatically for all the forbidden states, namely α, γ, ϵ, η of Fig. 21.

This general formulation of the exclusion principle was given by Dirac (1926). It has been found to be in perfect agreement with all relevant facts.

The role of the spin functions was to show how the symmetry of the orbital wave functions is connected with the total spin and to arrive at a proper formulation of the Pauli principle. The spin functions will only occur in this connexion.

VI

PERTURBATION THEORY

1. General theory

IF the interaction of two electrons is neglected, we have seen that the energy is the sum of the individual energies of the two electrons in states a and b:

$$E = E_a + E_b.$$

The question now arises as to how the total energy changes if we take into account the mutual interaction energy,

$$V_{12} = +\frac{e^2}{r_{12}}.$$

If V_{12} is inserted into the wave equation, the latter becomes far too complicated to be solved exactly. We can only look for suitable approximations. Hence we shall use a rather crude approximate method which, however, is sufficient for most problems and gives us a proper insight into what really happens. Better methods would have to be used if we wished to calculate atomic energy-levels with great accuracy, say of 1 or 2 per cent. But we shall not be concerned with such precision work here. The method we shall use can best be understood if we consider the problem in classical mechanics (and was actually developed first for astronomical purposes). Consider two particles moving on two different orbits (Fig. 22). In the absence of the interaction V_{12} the two particles move on the 'unperturbed orbits' (full curves). The interaction V_{12} causes the two orbits to be somewhat modified (dotted curves), but if the perturbation V_{12} is not very large the

PERTURBATION THEORY

modification of the orbits will not be very large either. At each position of the particles they have an interaction energy $V = e^2/r_{12}$. If we wish to know the average change of energy of the whole system due to V we have to average V over both orbits. Now if the perturbation of the orbits themselves, which—also due to V—is small, it is clear that, for a first approximation, we can form the average value of V by neglecting the change of the orbits themselves. The first approximation to the change of energy is therefore obtained by assuming that the particles still move on their *unperturbed orbits*, as if no mutual interaction existed. The change of energy of the whole system, due to the fact that they have an interaction energy V, is given by the average value $\bar{V}$, *averaged over the unperturbed orbits*.

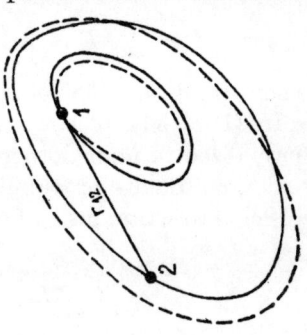

FIG. 22. Unperturbed and perturbed orbits of two particles.

The method can easily be applied to wave mechanics. Instead of well-defined orbits we have to deal with probability functions giving the probability for the position of the two particles. What correspond to the unperturbed orbits of classical theory are now the unperturbed wave functions of two electrons.

The two electrons in, say, the He atom have an interaction

$$V = +\frac{e^2}{r_{12}}.$$

To form the average perturbation energy we have to multiply V_{12} by the probability of finding the two particles at two specified positions, i.e. by $\psi^2(1, 2)$, which is the probability of finding the electrons at positions 1 and 2. For $\psi(1, 2)$

we may, as a first approximation, use the *unperturbed wave functions*, i.e. those calculated as if there were no mutual perturbation. The average value is obtained by integrating over all possible positions of the two electrons. Thus the average change of energy becomes

$$\Delta E = \int \psi^2(1,2) V \, d\tau_1 d\tau_2, \tag{1}$$

where $d\tau_1$, $d\tau_2$ are the volume elements of the two electrons.

In (1) it has, of course, been assumed that $\psi(1,2)$ is normalized to unity, otherwise its square does not represent the absolute probability. If ψ is not normalized yet we have to write, instead of (1),

$$\Delta E = \frac{\int V \psi^2 \, d\tau}{\int \psi^2 \, d\tau}, \tag{2}$$

writing $d\tau$ for $d\tau_1 d\tau_2$.

ΔE gives the first approximation of the energy change of the whole system due to the perturbation V. (2) is quite general and can be used for any kind of perturbation.

2. He atom, exchange energy

As an application of our general theory we consider the case of the two electrons in the He atom. To obtain the change of energy due to the perturbation $V_{12} = e^2/r_{12}$ we take the square of the wave function and insert it in equation (2). If the two electrons are in different unperturbed states a and b, the wave function is given by Chapter V, equation (13). It is essential that the symmetrical or antisymmetrical combinations are used, and not the product wave functions $\psi_a(1)\psi_b(2)$. Hence

$$\psi^2(1,2) = \psi_a^2(1)\psi_b^2(2) + \psi_b^2(1)\psi_a^2(2) \pm 2\psi_a(1)\psi_b(1)\psi_a(2)\psi_b(2).$$

The perturbation e^2/r_{12} is symmetrical in the two electrons 1 and 2. When integrated over the coordinates of both electrons the first two terms of (3) give equal contributions. Hence, cancelling the factors 2, we obtain

$$\Delta E = \frac{\int \{\psi_a^2(1)\psi_b^2(2) \pm \psi_a(1)\psi_b(1)\psi_a(2)\psi_b(2)\}\dfrac{e^2}{r_{12}}d\tau_1 d\tau_2}{\int \{\psi_a^2(1)\psi_b^2(2) \pm \psi_a(1)\psi_b(1)\psi_a(2)\psi_b(2)\} d\tau_1 d\tau_2}. \quad (4)$$

We shall not work out ΔE numerically (which can be done by inserting the wave functions ψ_a and ψ_b in (4)). Instead we confine ourselves to a qualitative discussion. Consider the first integral of equation (4) in the numerator and call it C:

$$C = \int \psi_a^2(1)\psi_b^2(2)\frac{e^2}{r_{12}} d\tau_1 d\tau_2.$$

ψ_a^2 is the probability of finding the electron that is in the quantum state a at a certain point, or the average charge density of the state a. Similarly, ψ_b^2 is the average charge density of the electron in the state b. The two charge distributions repel each other according to their Coulomb interaction. The *potential energy of this mutual interaction is just the integral* C, which has thus a simple natural meaning.

This Coulomb interaction of the two charge clouds C is, however, not the only contribution to the change of energy. The second term of the numerator of (4), which we denote by A, is

$$A = \int \psi_a(1)\psi_a(2)\psi_b(1)\psi_b(2)\frac{e^2}{r_{12}} d\tau_1 d\tau_2. \quad (5)$$

A cannot be so simply interpreted. It arises obviously from the fact that the two electrons are identical and can be exchanged. The electron 1 is 'partly' in the state a and partly in the state b $(\psi_a(1)\psi_b(1))$. If we wished we could interpret A also as the Coulomb interaction of two charge densities, or rather of different parts of one charge density with each other. This charge density would then be $\psi_a(1)\psi_b(1)$. (5) represents the Coulomb interaction of each part of the charge density $\psi_a(1)\psi_b(1)$ with every other part $\psi_a(2)\psi_b(2)$. We call it the *exchange charge density*. A will be called the *exchange energy*.

It must be remembered though that the interaction of this exchange charge with itself, namely A, occurs with both signs $+A$ and $-A$, and this fact cannot be explained by such a semi-classical picture of interacting charge clouds. The very occurrence of the exchange charge and the exchange energy A is a typical result of *quantum mechanics*. It is due to the fact that two electrons are *indistinguishable from each other*, and besides, of course, are described by wave functions. We have to accept the existence of such an exchange energy as a rather fundamental consequence of wave mechanics which cannot be explained on grounds of a classical picture. We shall discuss it in more detail in Chapter IX at the end of section 1.

The 'exchange energy' A is of great importance indeed for the understanding of atomic spectra, and more still for the understanding of chemistry. The *homopolar chemical bond* between two atoms rests on such an exchange energy (Chapter IX).

As to the denominator of (4) we may assume that ψ_a and ψ_b are normalized, $\int \psi_a^2(1)\, d\tau_1 = \int \psi^2(2)\, d\tau_2 = 1$. We shall furthermore show in section 3 that

$$\int \psi_a(1)\psi_b(1)\, d\tau_1 = 0.$$

Therefore the second integral in the denominator is equal to 0, and the denominator is equal to 1.

For a discussion of (4) we further remark that all the factors occurring in C are positive. Therefore C is positive. The same is true for A, although this cannot be shown so easily. If a is the ground state, it is true that ψ_a is always positive, but for the excited state ψ_b has positive and negative values (cf. Chapter III, Fig. 10). e^2/r_{12} is always positive. The largest contributions to the integral arise from small values of r_{12}, i.e. from two parts of the charge density $\psi_a\psi_b$ with the coordinates of the two electrons not very different from each other. $\psi_a(2)\psi_b(2)$ will then nearly always have the same sign as $\psi_a(1)\psi_b(1)$, and whatever the sign is, the integrand will be mostly positive. We therefore understand that the whole integral A is positive. The change of energy ΔE is now

$$\Delta E = C \pm A, \qquad (6)$$

with C and A both positive quantities.

A occurs with two different signs, and we obtain therefore *two different values* for the perturbation energy. We remember from Chapter V that the two signs of the unperturbed wave function

$$\psi_a(1)\psi_b(2) \pm \psi_a(2)\psi_b(1)$$

are attributed to the different values of the total spin $s = 0$ and 1 respectively. Therefore the singlet state has the perturbation energy

$$\Delta E = C+A \quad (s=0) \qquad (7a)$$

and the triplet the energy

$$\Delta E = C-A \quad (s=1). \qquad (7b)$$

If we have, for instance, a He atom in which one electron is in the ground state a and one electron in an excited state b, the unperturbed energy is for both triplet and singlet E_a+E_b. If we add the energy due to the perturbation ΔE, the triplet state will have a lower energy than the singlet state. Thus the unperturbed state splits up into two states with different energies, the energy difference being $2A$. The *exchange degeneracy is now removed*. A is by no means a small quantity. It depends on the *Coulomb interaction* e^2/r_{12} of the two electrons which is on the average, though, smaller than the potential energy of each electron but of the same order of magnitude. *The singlet-triplet splitting is therefore of the order of magnitude of the distance of the atomic levels themselves*. We can express this also in the following way: The *coupling of two electron spins* is accompanied by a *large coupling energy*, depending on whether the two spins are parallel or antiparallel. This coupling energy is of the order of magnitude of the electrical forces and by no means, as one might think, of the very small order of magnitude of the magnetic interaction of the spins. The reason is that, through the Pauli principle, the spins are tied to the symmetry of the orbital wave function.

If both electrons are in the same unperturbed level a, say, the wave function is $\psi_a(1)\psi_a(2)$ and the change of energy simply

$$\Delta E = \int \psi_a^2(1)\psi_a^2(2) \frac{e^2}{r_{12}} d\tau_1 d\tau_2. \qquad (8)$$

ΔE has only one value in accordance with the fact that only a singlet state exists, by the Pauli exclusion principle.

The change of energy due to the perturbation of the two electrons in the He atom is shown qualitatively in Fig. 23, for the ground state and the first excited state.

The values of the integrals have been worked out numerically, and the energy-levels have been found to agree within 20 or 30 per cent. with the observed ones.

More we cannot expect from such a crude perturbation method. Especially, it has been found that the first excited state of He is a triplet, and this is precisely what Fig. 23 shows to be the case.

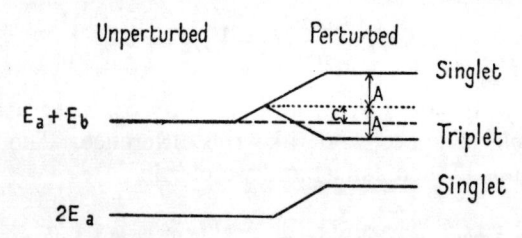

FIG. 23. Perturbation energy of He levels.

With the help of more refined methods the He states can be calculated with much greater precision. In particular the ground state has been worked out by Hylleraas (and others) with an error of less than 0·0001 per cent. The perfect agreement of this result with the experimental value is a very valuable test of the theory and shows that the many-body problem is also described correctly by wave mechanics.

3. The orthogonality theorem

In the preceding section we have used the fact that

$$\int \psi_a(1)\psi_b(1)\, d\tau_1 = 0. \qquad (9)$$

(9) is quite a general theorem, called the orthogonality relation, which we have now to prove. It refers to the quantum states of a single electron and holds for any two different states and in whatever potential field the electron moves.

PERTURBATION THEORY

We write down the wave equation for the two quantum states a and b:

$$\nabla^2 \psi_a + \frac{2m}{\hbar^2}(E_a - V)\psi_a = 0, \qquad (10a)$$

$$\nabla^2 \psi_b + \frac{2m}{\hbar^2}(E_b - V)\psi_b = 0. \qquad (10b)$$

We multiply $(10a)$ by ψ_b and $(10b)$ by ψ_a, integrate over the whole of space, and take the difference. The terms containing V then cancel and we obtain

$$\int (\psi_b \nabla^2 \psi_a - \psi_a \nabla^2 \psi_b)\, d\tau = -\frac{2m}{\hbar^2}(E_a - E_b)\int \psi_a \psi_b\, d\tau. \qquad (11)$$

Consider the left-hand side. $\nabla^2 = \frac{\partial^2}{\partial x^2} + \frac{\partial^2}{\partial y^2} + \frac{\partial^2}{\partial z^2}$. The integration over $d\tau$ includes an integration over x from $-\infty$ to $+\infty$. We take the first term $\partial^2/\partial x^2$ and integrate over x. Integrating by parts,

$$\int_{-\infty}^{+\infty}\left(\psi_b \frac{\partial^2}{\partial x^2}\psi_a - \psi_a \frac{\partial^2}{\partial x^2}\psi_b\right)dx$$

$$= \left|\left(\psi_b \frac{\partial}{\partial x}\psi_a - \psi_a \frac{\partial}{\partial x}\psi_b\right)\right|_{-\infty}^{+\infty} - \int_{-\infty}^{+\infty}\left(\frac{\partial \psi_b}{\partial x}\frac{\partial \psi_a}{\partial x} - \frac{\partial \psi_a}{\partial x}\frac{\partial \psi_b}{\partial x}\right)dx. \qquad (12)$$

The integral on the right is obviously zero. The first term on the right has to be taken at an infinitely large distance $+\infty$ or $-\infty$, but here the wave function must vanish. This is indeed the 'boundary condition' which we had to impose on all wave functions (Chapter II). (12) therefore vanishes. The same is true for the parts $\partial^2/\partial y^2$ (integrating

PERTURBATION THEORY 95

over y) and $\partial^2/\partial z^2$. Hence the left-hand side of (11) vanishes. We therefore find that always

$$(E_a - E_b) \int \psi_a \psi_b \, d\tau = 0. \qquad (13)$$

If now a and b are states with two different energies $E_a \neq E_b$ the integral of (13) must vanish, which proves our statement (9).

It may be that ψ_a and ψ_b are two different wave functions belonging to the same energy. In this case we are dealing with a degenerate state. The relation (9) need then not necessarily hold. On the other hand, if ψ_a, ψ_b are two degenerate wave functions, we are free to choose any two linear combinations of ψ_a and ψ_b, and this fact can be used to *make* the two new wave functions orthogonal (i.e. to choose such combinations that (9) holds). Suppose, for instance, that the integral (9) has the value c. We then choose, instead of ψ_a and ψ_b,

$$\psi'_a = \psi_a, \qquad \psi'_b = \psi_b - c\psi_a.$$

If ψ_a was normalized $\left(\int \psi_a^2 \, d\tau = 1 \right)$, we find

$$\int \psi'_a \psi'_b \, d\tau = c - c \int \psi_a^2 \, d\tau = 0.$$

We can therefore assume that (9) holds for any two different wave functions, whether they belong to two different energies or to the same energy. The reader will easily satisfy himself that the three p wave functions ψ_x, ψ_y, ψ_z (Chapter III) are orthogonal. The same is true for the combinations ψ_{+1}, ψ_0, ψ_{-1}. They are, of course, also orthogonal to the wave function of the ground state of hydrogen.

In Chapter V, section 5, we had also introduced spin wave functions α and β which describe the spin of one electron in the two states with spin component in the z

direction $m_{sz} = \pm\tfrac{1}{2}$. Also these two spin functions must be orthogonal to each other because they belong to two different states which have different energies in a magnetic field:

$$\int \alpha\beta = 0. \tag{14}$$

The integration is to be carried out over all values of the variable (or variables) on which α and β depend. (In Chapter V, section 5, we had provisionally introduced some angle of rotation for this variable but this is irrelevant.) Also α and β may be assumed to be normalized

$$\int \alpha^2 = \int \beta^2 = 1. \tag{15}$$

VII

THE PERIODIC SYSTEM OF ELEMENTS

1. The electron configuration

THE atoms of the different chemical elements consist of a nucleus of charge ze, surrounded by z electrons, where z is an integer ranging from 1 (hydrogen) to 92 (uranium). If we neglect the interaction between the electrons each electron moves independently in a potential $V = -ze^2/r$. This is the same potential as in a H atom, only that e^2 is replaced by ze^2. The arrangement of the energy-levels for the individual electrons is therefore the same for all atoms and especially the same as for hydrogen. Only the scale of energy is different. We see that at once from formula (7), Chapter III, giving the energies of the hydrogen levels. We obtain the levels for the individual electrons of any other atom by replacing e^4 by z^2e^4. Apart from the scale the arrangement of the one-electron levels is therefore always that of Fig. 11. To find the levels for the whole atom we distribute all the z available electrons amongst the energy-levels of Fig. 11, in all possible ways.

In doing so we must remember, however, that not all distributions are possible. The exclusion principle (Chapter V) demands that in each one-electron level not more than two electrons can be placed (and they must have, then, antiparallel spin).

The different one-electron levels are classified by the angular momentum, e.g. $l = 0, 1, 2,...$, etc. We call them $s, p, d,...$ levels respectively. If the angular momentum is l, then the level is $(2l+1)$-fold degenerate and actually consists of $2l+1$ coinciding levels. In such a level therefore $2(2l+1)$ electrons can be placed.

The order of the levels, arranged according to increasing energies, is $1s$, $2s$, $2p$, $3s$, $3p$, $3d$, $4s$, etc., where $1s$, $2s$,... denote the lowest, the second, etc., s levels.

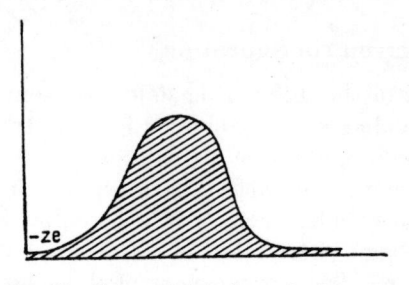

FIG. 24. Charge cloud of two $1s$ electrons.

In the H atom the $2s$ and $2p$ levels coincide, the same is true for the higher $3s$, $3p$, $3d$ levels. This is rather an accidental fact and is only true for a Coulomb potential ze^2/r. In fact the s and p levels are somewhat separated except for hydrogen itself, for the following reason: Consider the case of three electrons with two electrons in the $1s$ level and one in the $2s$ or $2p$ level. It would now be a very poor approximation if we were to assume that the electron in the higher level moves in the pure field of the nucleus $-ze^2/r$. The two electrons in the lower level, which are, according to Fig. 10, p. 38, much nearer to the nucleus than the $2s$ electron, represent a negative charge cloud, roughly, as shown in Fig. 24. This charge cloud screens the Coulomb field to some extent. At a very large distance the potential is only $-(z-2)e^2/r$. At medium distances the potential is between $-ze^2/r$ and $-(z-2)e^2/r$. Actually, therefore, the electron in the higher level does not move in an exact Coulomb field, the 'effective charge' of the nucleus depending on the distance and ranging from z to $z-2$.

The effect is, that the $2s$ and $2p$ levels are more or less separated. The same is true, of course, for the $3s$, $3p$, $3d$ levels, the separation is bigger for the higher elements where z is large. Actually, the separation is such that for light elements it is always the s level which is the lowest one, the p, d levels lying higher, which cannot, however, be shown here in detail. The separation of the $2s$ and $2p$ levels is, though quite appreciable, by no means so large as that of the $1s$ and $2s$ levels or between $2s$ and $3s$.

We are chiefly interested in the lowest state which is the normal state of an atom. We obtain it by filling up gradually all the lowest one-electron levels with the maximum number of electrons permitted by the exclusion principle. Take, for instance, the case of the Li atom with three electrons. The first two electrons can be placed in the $1s$ level. The third electron is then in the next higher level, which is $2s$. Next comes beryllium with four electrons, of which two are placed in $1s$ and two in $2s$. For the heavier elements the $2p$ level is gradually filled up. It is completed by $2.3 = 6$ electrons. This is the case for Ne, which has ten electrons, two in $1s$, two in $2s$, and six in $2p$. When a level is filled completely we call it a 'closed shell'. Hence the first closed shell is completed for helium. Since the separation of the $2s$ and $2p$ levels is not very large, one usually considers them together and speaks of the next closed shell only when both are completed, which is first the case for Ne. In this way we find the electron distribution for all the elements. For the first two periods of the periodic system it is shown in Table 6.

The chemical behaviour of an atom depends on those electrons which are not parts of a closed shell. The chemical valency is in fact either equal to the number of free electrons not contained in a shell or to the number of electrons missing to complete a shell, whichever is the smaller of the two numbers. Elements which differ only by a closed shell and have otherwise similar electron configurations behave chemically very much alike. From Table 6 it is clear that the chemical behaviour must be

TABLE 6. *Electron distribution in periodic table*

	H	He	Li	Be	B	C	N	O	F	Ne
z	1	2	3	4	5	6	7	8	9	10
$1s$	1	2	2	2	2	2	2	2	2	2
$2s$			1	2	2	2	2	2	2	2
$2p$					1	2	3	4	5	6

			Na	Ca	Al	Si	P	S	Cl	A
z			11	12	13	14	15	16	17	18
$1s$			2	2	2	2	2	2	2	2
$2s$			2	2	2	2	2	2	2	2
$2p$			6	6	6	6	6	6	6	6
$3s$			1	2	2	2	2	2	2	2
$3p$					1	2	3	4	5	6

periodic. For instance, H, Li, Na have, apart from closed shells, one free electron. F, Cl,... have one electron missing to complete a shell. The elements which have only closed shells, namely, He, Ne, A,... are the rare gases.

In this way quantum theory accounts for the well-known periodic system of chemical elements.

2. The atomic states

We shall now deal in more detail with the first period of the periodic table. The chemical behaviour of an atom depends not only on the electron configuration but also on further details such as the spin directions of the free electrons, etc. It is especially the spin directions of the

THE PERIODIC SYSTEM OF ELEMENTS 101

extra-shell electrons in the lowest state of the atom which are of importance. We shall see that, in general, out of each electronic configuration several atomic states arise which are distinguished by the total spin and angular momentum. Owing to the interaction between the electrons (which has so far been neglected) these states will have different energies and we have to find out which of them is the lowest state. The various states arising from one electron configuration can be found by the vector addition rules of Chapter IV.

We discuss now all the elements of the first period: The H atom has one electron in the $1s$ level. The spin is $\frac{1}{2}$, therefore the lowest level is a ^2S state. Helium has two electrons in the $1s$ level. They must have antiparallel spin (Pauli principle), therefore the ground state is ^1S. In the Li atom the third electron is in the first excited $2s$ level with spin $= \frac{1}{2}$. Since the two electrons of the $1s$ shell do not contribute to the spin the ground state is ^2S. All the alkalis have a ^2S as ground state since the closed shells do not contribute to the spin. For beryllium the electron configuration is similar to that of helium with a ^1S as ground state. Boron has one electron more in the $2p$ level. The total spin is therefore $s = \frac{1}{2}$. In addition, the p electron has an angular momentum $l = 1$. The atomic state is therefore ^2P. For the five elements H, He, Li, Be, B there is only one atomic state arising from the lowest electron configuration. Their ground states are represented in Fig. 25.

From carbon onwards the atomic states are more complicated. C has two electrons in the $2p$ shell in addition to the two electrons in $2s$. We denote such a configuration by s^2p^2. The p level really consists of three levels with $m_z = +1, 0, -1$. The two electrons can be distributed in various ways amongst these three levels. In addition, the two electrons can have parallel or antiparallel spin directions. Only, of course, if they are both in the same level, the spins must be antiparallel because of the exclusion principle.

FIG. 25. Ground states of H,..., B.

The various distributions of the two p electrons are shown in Fig. 26. Altogether there are nine different configurations. Each configuration has a total spin $s = 1$ or 0, according to whether the two spins are parallel or anti-

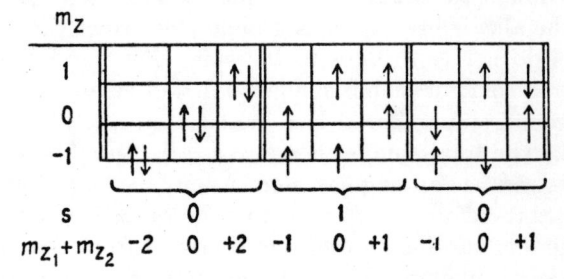

FIG. 26. Electron distribution in C.

parallel, and belongs therefore to a triplet or singlet state respectively. In addition, each configuration has an orbital angular momentum the z component of which is simply the sum of the contributions of the two electrons. It is given in the bottom line. To find the atomic states we remember that each atomic state has an orbital momen-

THE PERIODIC SYSTEM OF ELEMENTS 103

tum l with components $m_z = -l, -l+1,..., +l$. We therefore have to group together those configurations with different values m_z belonging to the same value of l.

Collecting first the configurations with $s = 1$ we have $m_z = -1, 0, +1$. These three configurations therefore represent a state, with $l = 1$, $s = 1$, i.e. a ^3P state. For $s = 0$ we have $m_z = -2, -1, 0, +1, +2$, and one more 0. They belong to two atomic states with $l = 2$ and $l = 0$. We therefore obtain ^1D and ^1S. Altogether we have three atomic states arising from the two p electrons, namely ^3P, ^1D, ^1S. If the interaction between the electrons is taken into account, all these three states have different energies. We can easily find the ground state. We remember from the perturbation theory, Chapter VI, that the energy is lowest when the spins are parallel. Therefore we conclude that the ground state of the carbon atom is ^3P, the ^1D and ^1S being excited states. This is also in agreement with the spectroscopical facts.

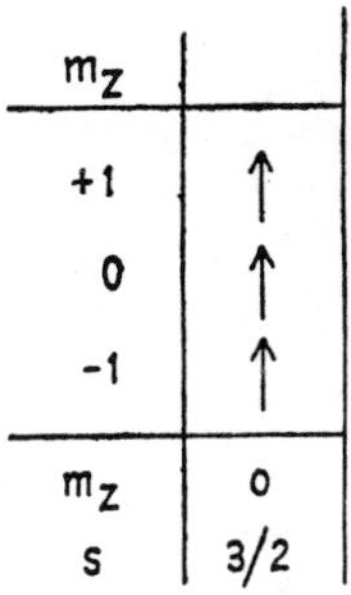

m_z	
+1	↑
0	↑
-1	↑
m_z	0
s	3/2

FIG. 27. Ground state of N.

The next element is nitrogen, which has three electrons in the p shell (configuration s^2p^3). If we are only interested in the ground state we need only consider those configurations where all three electrons have parallel spin. There must then be one electron in each of the three levels. There is only one such configuration. Here $m_z = 0$, therefore $l = 0$ (Fig. 27). Since $s = \frac{3}{2}$, the state is a ^4S state. This is in fact the ground state of N. In addition, there are also excited states arising from the same electron configuration. They are in fact ^2D and ^2P which we leave to the reader to show.

The remaining three elements, O, F, Ne are easily

104   THE PERIODIC SYSTEM OF ELEMENTS

discussed. We confine ourselves to the ground state, i.e. the state with as many electrons with parallel spin as possible. The electron configurations for the three elements in this case are shown in Fig. 28. The ground states are therefore: ^3P for O (s^2p^4), ^2P for F (s^2p^5), and ^1S for Ne (s^2p^6).

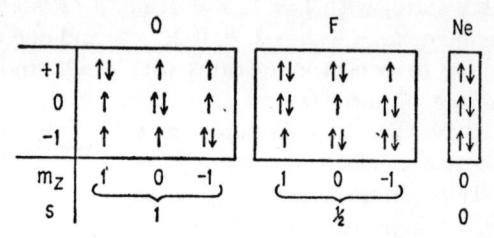

	O	F	Ne
+1	↑↓ ↑ ↑	↑↓ ↑↓ ↑	↑↓
0	↑ ↑↓ ↑	↑↓ ↑ ↑↓	↑↓
−1	↑ ↑ ↑↓	↑ ↑↓ ↑↓	↑↓
m_z	1 0 −1	1 0 −1	0
s	1	½	0

FIG. 28. Ground states of O, F, Ne.

There are also excited states of O, arising out of the same electron distribution. They are ^1D and ^1S. F and Ne have no excited states from the lowest electron configuration.

Finally we give a table of the ground states of all the elements of the first period (Table 7):

TABLE 7. *Ground states of the first period of elements*

H	He	Li	Be	B	C	N	O	F	Ne
^2S	^1S	^2S	^1S	^2P	^3P	^4S	^3P	^2P	^1S

We may also consider further excited states by considering higher electron configurations. For carbon, for instance, we can raise one of the 2s electrons into the 2p level. There are then three electrons in 2p and one in 2s (configuration sp^3). All these four electrons can have parallel spin (Fig. 29). This will be a quintet state ($s = 2$) and since $l = 0$ (the three p electrons have a total orbital momentum zero as in the case of N) the state is ^5S. This ^5S state of C is of great importance for the chemical behaviour of carbon, as it is responsible for the four valencies of C. This question will be dealt with in Chapter X.

The energies of the atomic levels are mainly determined by three factors: (i) the electron configuration from which

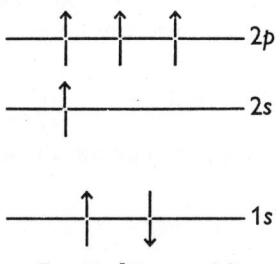

FIG. 29. ^5S state of C.

the level arises; (ii) the screening of the inner electrons which causes the s-p splitting; (iii) the interaction of the electrons which causes a splitting of the various terms of the same electron configuration (neglecting the very small fine structure splitting). Although these effects are all of the same order of magnitude they become, on the whole, decreasingly smaller in the order they were mentioned. To illustrate this we give in Table 8 the excitation energies of some of the levels of carbon as they are known from spectroscopy.

TABLE 8. *Excitation energies of carbon levels*

Levels Number of electrons	2s 2	2p 2		2s 1	2p 3	2s 2	2p 1	3s 1
Atomic term	^3P	^1D	^1S	^5S		^3P		^1P
Excitation energy (eV)	0	1·3	2·7	4·2		7·4		7·7

The terms of the lowest configuration are only separated by 2 or 3 electron volts. Raising one 2s electron into the

$2p$ level requires at least 4 volts. There are many more terms arising from this configuration, apart from the ^5S (not given in the table), which lie much higher. The raising of a $2p$ electron into the next higher level $3s$ requires much more energy, 7–8 eV. The energy required to remove one $2p$ electron altogether, i.e. the ionization energy, is 11·2 eV Of all the terms arising from the same electron configuration it is always the term with the highest multiplicity which is the lowest, in agreement with what we found in Chapter VI.

The situation is very similar for all the light elements.

VIII

DIATOMIC MOLECULES

1. The electronic states

IN this chapter we consider the quantum states of a diatomic molecule consisting of two nuclei and a number of electrons. In general all these particles are in motion and interact with each other. At first sight the problem of finding quantum states looks rather formidable, but we are more concerned with a general survey than with an exact calculation of levels and this is obtained quite easily. The chief clue is the fact that the *nuclei* are very much *heavier* than the electrons. They move, therefore, much more slowly. If they were at rest and in fixed positions, at a distance R say from each other, the problem would be reduced to finding the energy levels of the *electrons* moving in a potential created by the two centres of force, but the motion of the nuclei would not enter into the problem. Now when the nuclei move but slowly this cannot create a drastically new situation. For each instant of time t we may, approximately, regard the nuclei as if they were at rest and determine the energy of the electrons for the particular distance $R(t)$ which the nuclei have at the time. In this way we obtain *electronic energy levels* $E_n(R)$ which all depend on a parameter R, the distance between the nuclei. The problem is thus split up into two parts, (i) the determination of the electronic levels, and (ii) the motion of the nuclei themselves. The latter problem will be considered in sections 2 and 3. It is true that this splitting up of the problem is not exact (because the motion of the nuclei influences that of the electrons and vice versa) but it is a good approximation and quite sufficient for our purpose.

We can easily obtain a qualitative idea of the R-dependence of at least the low energy levels. When the molecule is in equilibrium, i.e. in a state of minimum energy, the nuclei are at a finite distance R_0. The lowest electronic level $E_1(R)$ has therefore a minimum energy at R_0. For larger and smaller R, E_1 will increase. It is convenient to include in $E(R)$ also the Coulomb interaction between the two nuclei which is always positive (repulsion) because the nuclei are both positively charged. Thus at small distances (and this also applies to all excited states) $E_1(R)$ increases rapidly and becomes positive. On the other hand, for $R > R_0$, E_1 also increases. When R is very large, the energy is equal to that of the two separated atoms. The lowest electronic molecular state will naturally go over asymptotically into two atoms in their ground state. The R-dependence of the excited molecular states $E_n(R)$ will be similar. It happens frequently that different molecular states go over asymptotically into the same states of the separated atoms or, in other words, when we let two atoms in definite states interact with each other and let their distance decrease, several types of molecular states may arise (see below). Thus we obtain the type of molecular states or 'atomic interaction curves' shown in Fig. 30. $E(\infty) - E(R_0)$ is the dissociation energy.

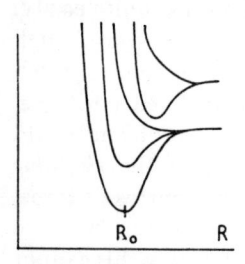

FIG. 30. Atomic interaction curves.

In addition to the above type of states which lead to stable molecules there are states where the electronic energy is always larger than that of the separated atoms. (Two such curves are also shown in Fig. 30.) An example is derived in Chapter IX, p. 133. These are not proper molecular states because the atoms then repel each other and would at once dissociate. Such repulsive states are important when we consider the interaction between atoms in connexion with the chemical bond but need not be considered in this chapter.

We have seen in the preceding chapters that the states of an atom can be classified according to their total spins and their total angular momentum l. Accordingly, we speak of ^2S, ^3P, etc., states. A similar classification can also be carried out for molecules. Consider in the first place the orbital angular momentum. If a particle moves in a field of central symmetry (atom), then according to classical mechanics all three components of the angular momentum (and therefore the absolute value of the angular momentum) remain constant in course of time. In quantum theory this finds its expression in the fact that two quantum numbers, the total angular momentum l and its z component m_z, exist which remain constant in course of time and can therefore be used to characterize the atomic state, whilst the other two components m_y, m_x are not sharp. This is not quite so for a diatomic molecule. The electric field of such a molecule no longer has central symmetry but is only symmetrical with regard to rotations about the axis of the molecule. A particle, or several particles, moving in such a field of axial symmetry have, according to classical mechanics, but one component of angular momentum which remains constant in course of time, namely the angular momentum about the axis of the molecule. In quantum theory this must mean that a quantum number μ, independent of R, exists, which denotes the angular momentum about the molecular axis, and that μ like m_z can have only integral values. μ can also be negative because the particles in a molecule can rotate about the axis clockwise as well as anticlockwise. Now the energy of the molecule cannot depend on the sense of rotation. States with angular momentum μ and $-\mu$ will therefore have the same energy. The molecular states are thus always *two-fold degenerate*, except when $\mu = 0$. We can, in analogy to the atomic l, define the absolute value of the angular momentum $\lambda = |\mu|$ and classify the molecular states according to the value λ. Only when $\lambda = \mu = 0$ are the states not degenerate. We call, in analogy to the S, P, D,... states of an atom, the states with $\lambda = 0, 1,...$ Σ, Π, Δ,... states respectively. Thus we have the scheme

$$\begin{array}{cccc} & \Sigma & \Pi & \Delta & ... \\ \lambda = & 0 & 1 & 2 & ... \\ \mu = & 0 & 1, -1 & 2, -2 & ... \end{array}$$

Next consider the spin. The spin has practically no interaction with the orbital motion of the particles. It is able to retain its *free rotation*, no matter in what sort of field the particles move. The spin of a molecule has therefore exactly the same properties and follows the same addition rules as in an atom. There will be molecular states with spin 0, $\frac{1}{2}$, 1,... or singlet, doublet, triplet states, etc. Molecular states will therefore have notations such as $^1\Sigma$, $^2\Pi$, $^3\Pi$, $^1\Delta$,..., etc.

What states of a molecule arise, then, if two atoms, specified by their s and l values, are brought in contact? The spin values are obtained from the ordinary vector addition rules and require no further explanation. Let l_1 and l_2 be angular momenta of the two atoms and m_1 and m_2 the z components. $m_1 = -l_1,..., +l_1$, $m_2 = -l_2,..., +l_2$. Let also the axis of the molecule be the z-axis. The angular momentum about the molecular axis μ is obviously the sum $m_1 + m_2$. Thus a variety of μ-values arise by adding up the various m_1 and m_2. Those μ-values which differ only by their sign belong to the same λ and are part of a two-fold degenerate state. Different λ's belong to different

DIATOMIC MOLECULES

molecular states. Thus an atom in an S state and an atom in a P state give rise to two molecular states with $\lambda = 0$ and 1, i.e. to a Σ and a Π state. Both occur with all possible values of the spin obtained by the vector rule. We give two examples; a pair of atoms in ^2S and ^2P states respectively, and a pair of atoms both in ^3P states:

$^2S + {}^2P = {}^1\Sigma, {}^1\Pi, {}^3\Sigma, {}^3\Pi,$

$^3P + {}^3P = {}^1\Sigma, {}^1\Sigma, {}^1\Sigma, {}^1\Pi, {}^1\Pi, {}^1\Delta, {}^3\Sigma, {}^3\Sigma, {}^3\Sigma, {}^3\Pi, {}^3\Pi, {}^3\Delta,$
$\phantom{^3P + {}^3P = }{}^5\Sigma, {}^5\Sigma, {}^5\Sigma, {}^5\Pi, {}^5\Pi, {}^5\Delta.$

All these states have different energies when the atoms interact. Not all of them are states of a stable molecule. Some, in fact the majority, correspond to repulsion of the two atoms. In the case of two H atoms both in ^2S the two states arising are $^1\Sigma$ and $^3\Sigma$; the $^1\Sigma$ is the molecule and the $^3\Sigma$ will be seen to be a state where the two atoms repel each other (Chapter IX).

We comprise the characteristics of a molecular state, namely spin and angular momentum, by one word, 'race'. We say that two states belong to the same race if they have the same spin and the same angular momentum, but to a different race if one or both of them are different. Two $^1\Sigma$ states belong to the same race, but a $^1\Sigma$ and $^1\Pi$ to different races.†

Thus when two atoms in specified states are brought into contact several modes of interaction are possible, according to which of the above states are formed. In some cases a stable molecule is formed (in one of its electronic states) and in some cases the atoms will repel each other.

† Actually there are still more characteristics which define a molecular state, namely certain reflection properties of the wave function. We shall not use them in the following, and refer the reader to special books on molecules about this point.

2. The rotation of molecules

In the preceding section we have seen that the electronic levels of a molecule can be calculated by assuming the nuclei to be in fixed positions not necessarily at the equilibrium distance. We now consider the motion of the nuclei. Let us assume that the electrons are, for example, in their lowest electronic state, with energy $E_1(R)$ represented by the lowest curve of Fig. 30. Now the nuclei are allowed to move in space. If their distance R changes the electronic energy changes according to the function $E_1(R)$. This change must now govern the motion of the nuclei. In other words $E_1(R)$ will play the role of a *potential energy* under the influence of which the nuclei will move.

Let us first consider the motion of the nuclei, according to classical mechanics. It will be convenient to introduce the centre of gravity of the two nuclei† as the origin of our frame of reference and we may well assume that this is at rest. As is well known the centre of gravity could only move with uniform velocity and this is not an interesting motion. If m_1, m_2 are the masses, $\mathbf{r}_1$, $\mathbf{r}_2$ the position-vectors of the two nuclei with the centre of gravity as origin, then

$$m_1 \mathbf{r}_1 = -m_2 \mathbf{r}_2. \tag{1}$$

If we use this relation and $\mathbf{v}_1 = d\mathbf{r}_1/dt$ the kinetic energy of the nuclei becomes

$$T = \frac{m_1}{2}\mathbf{v}_1^2 + \frac{m_2}{2}\mathbf{v}_2^2 = \tfrac{1}{2}m(\mathbf{v}_1-\mathbf{v}_2)^2, \quad m = \frac{m_1 m_2}{m_1+m_2}. \tag{2}$$

m is the reduced mass. If we introduce the distance between the nuclei

$$\mathbf{R} = \mathbf{r}_1 - \mathbf{r}_2 \quad \text{and} \quad \mathbf{v} = \frac{d\mathbf{R}}{dt} = \mathbf{v}_1 - \mathbf{v}_2$$

---

† Since the electrons are very light compared with the nuclei, the centre of gravity of the molecule is practically identical with the centre of gravity of the two nuclei.

DIATOMIC MOLECULES

the total energy is
$$H = \tfrac{1}{2}m\mathbf{v}^2 + E_1(R). \tag{3}$$

This is the energy function for the motion of the nuclei and it is seen that it depends only on one variable, namely $\mathbf{R}$, and its time derivative. H is, of course, constant, $H \equiv E$. Consider now an energy value E higher than the minimum of the 'potential' E_1, i.e. higher than $E_1(R_0)$, but lower than the electronic energy of the separated atoms ($E_1(\infty)$). The absolute value of the distance R can then vary between two extreme values, R_1 and R_2 say, given by the two branches of the potential curve Fig. 30. In addition the direction of $\mathbf{R}$ can change freely in space. If R were constant, this would be a free *rotation*. Thus we are dealing with a vibration (change of R between two extreme values), as well as a rotation round the centre of gravity. To see this more clearly we put the kinetic energy into a different form. For this purpose we introduce the angular momentum of the two nuclei round the centre of gravity

$$\mathbf{M} = m_1(\mathbf{r}_1 \times \mathbf{v}_1) + m_2(\mathbf{r}_2 \times \mathbf{v}_2) = \frac{m_1}{m_2}(m_1 + m_2)(\mathbf{r}_1 \times \mathbf{v}_1)$$
$$= m(\mathbf{R} \times \mathbf{v}). \tag{4}$$

The square is given by
$$\mathbf{M}^2 = m^2(\mathbf{R}^2\mathbf{v}^2 - (\mathbf{R}\mathbf{v})^2). \tag{5}$$

Furthermore, the change with time of the absolute value of the distance R is given by

$$\frac{dR}{dt} \equiv \frac{d}{dt}\sqrt{\mathbf{R}^2} = \frac{(\mathbf{R}\mathbf{v})}{R}. \tag{6}$$

Thus with the help of (5) and (6) we can express $\mathbf{v}^2$ in terms of $\mathbf{M}^2$, $(dR/dt)^2$ and R:

$$\mathbf{v}^2 = \frac{\mathbf{M}^2}{m^2 R^2} + \left(\frac{dR}{dt}\right)^2,$$

and the energy of the nuclei becomes

$$H = \frac{\mathbf{M}^2}{2mR^2} + \tfrac{1}{2}m\left(\frac{dR}{dt}\right)^2 + E_1(R). \tag{7}$$

Now in many cases the extreme values R_1 and R_2 between which R can vary, do not differ very much. This is especially so, if the energy is not much higher than the minimum $E_1(R_0)$. (In the minimum itself R equals R_0 and the two extreme values coincide.) In such cases we may well replace R in the first term by a constant, for example by the average $(R_1+R_2)/2$.† The energy then splits up into two contributions, $H = H_{\text{rot}} + H_{\text{vib}}$, a rotational energy

$$H_{\text{rot}} = \frac{\mathbf{M}^2}{2mR^2} \tag{8a}$$

and a vibrational energy

$$H_{\text{vib}} = \tfrac{1}{2}m\left(\frac{dR}{dt}\right)^2 + E_1(R). \tag{8b}$$

$mR^2 \equiv A = m_1 r_1^2 + m_2 r_2^2$ (by (1)) is the moment of inertia of the molecule.

Now also the motion of the nuclei has to be treated according to quantum mechanics. This is quite simple if the above separation into rotation and vibration is a feasible approximation. Consider first the rotational energy (8a). As we have put $R = $ const. here, this depends on the square of the angular momentum only. We have considered the angular momentum of an electron (in a central field of force) in detail in Chapter IV and we have seen that $\mathbf{M}^2$ has a *sharp value*

$$\mathbf{M}^2 = \hbar^2 l(l+1), \quad l = 0, 1, 2,... \tag{9}$$

† This, of course, cannot be done in the potential energy $E_1(R)$ because this varies very much between R_1 and R_2.

simultaneously with the total energy. The case of a rotating molecule could only differ by the mass of the particle, but this enters in the expression for the energy only, and not in the value of M^2. Thus (9) must hold for a molecule also. Thus H_{rot} itself (and therefore also $H_{\text{vib}} = E - H_{\text{rot}}$) has a sharp value but this fact depends on the approximation $R^2 \simeq \text{const.}$

$$H_{\text{rot}} \equiv E_{\text{rot}} = \frac{\hbar^2}{2A} l(l+1). \qquad (10)$$

We obtain a series of rotational energy levels for $l = 0, 1, 2, \ldots$. These energy values and their spacing are, of course, much smaller than for electrons because the (large) mass of the nuclei enters in the denominator. For the H_2 molecule, at the equilibrium distance ($R = 0.8$ Å.U.), we obtain $\hbar^2/2A = 7 \cdot 3 \times 10^{-3}$ eV whereas the electronic energies are of the order of a few eV. In the next section we consider the vibrational energy.

3. The vibration of molecules

The vibrational energy of a diatomic molecule is given by (8b). If we treat the vibration according to wave mechanics we have to replace the velocity by the momentum $P = m\,dR/dt$ and then replace P by the operator $-i\hbar\,\partial/\partial R$. It is a one-dimensional motion with the distance, R, between the atoms as variable. The wave equation is thus

$$\frac{d^2\psi}{dR^2} + \frac{2m}{\hbar^2}(E - E_1(R))\psi = 0. \qquad (11)$$

It is quite easy to obtain a qualitative idea of the energy levels. Looking at Fig. 30 we can say at once that all discrete levels E_n must lie between two energy limits: E_n must certainly be larger than the minimum of the potential, $E_1(R_0)$, and it must be smaller than the asymptotic value $E_1(\infty)$ of the atomic interaction curve. For if E is larger than $E_1(\infty)$ the atoms can move to infinity and we have a continuous spectrum just as in the case of the Coulomb potential. In fact the curves of Fig. 30 have this qualitative feature in common with the Coulomb potential that both tend to a finite value at large distances (compare Fig. 7, p. 24). It might even be conjectured that the energy levels crowd as E approaches the limit $E_1(\infty)$. This is in fact the case for most electronic states of diatomic molecules. Thus we obtain the qualitative level scheme shown in Fig. 31.

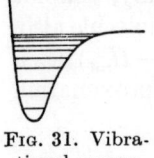

FIG. 31. Vibrational energy levels.

For the low levels a feasible approximation can be made. We may replace the curve Fig. 31 in the neighbourhood of the minimum by a parabola

$$E_1(R) \simeq f^2(R-R_0)^2 + E_1'(R_0); \qquad (12)$$

f is determined by the curvature of the curve near the minimum. Introducing then $R-R_0 = x$ as variable the wave equation becomes that of a harmonic oscillator

$$\frac{d^2\psi}{dx^2} + \frac{2m}{\hbar^2}(E' - f^2 x^2)\psi = 0, \qquad (13)$$

$$E' = E - E_1(R_0).$$

The solutions of (13) are not difficult to find. The reader may easily verify that the three lowest solutions are (not normalized)

$$\psi_1 = e^{-\alpha x^2}, \qquad \psi_2 = x e^{-\alpha x^2}, \qquad \psi_3 = (1-4\alpha x^2)e^{-\alpha x^2}, \qquad (14)$$

$$\alpha = \frac{f}{\hbar}\sqrt{\frac{m}{2}},$$

with the corresponding energies

$$E_1' = \tfrac{1}{2}\hbar\omega, \qquad E_2' = \tfrac{3}{2}\hbar\omega, \qquad E_3' = \tfrac{5}{2}\hbar\omega, \qquad \omega = f\sqrt{\frac{2}{m}}.$$

In fact the general formula for the energy levels is

$$E'_n = (n+\tfrac{1}{2})\hbar\omega. \tag{15}$$

The energy levels all have equal spacing. This is, of course, only true as long as the interaction curve can be approximated by a parabola.

The lowest level is higher than the minimum of the potential $E_1(R_0)$ by an amount $\tfrac{1}{2}\hbar\omega$. This 'zero point energy' is due to the uncertainty principle. To make the energy small we have to make the kinetic energy small. Now in order to minimize the kinetic energy (and hence the momentum) a certain extension in space is required, but at the minimum of the potential the position would be fixed at R_0. Therefore a certain possibility for vibration is needed which in turn means a certain increase of energy.

The spacing of the low vibrational levels is usually much larger than that of the low rotational levels. For the H_2 molecule, for example, $\hbar\omega$ is about $\tfrac{1}{2}$ eV. Thus each vibrational level is followed by a band of narrowly spaced rotational levels. The experimental analysis of the vibrational levels is an important tool for the determination of the actual atomic interaction curves.

4. Ortho- and para-molecules

If a diatomic molecule consists of two like atoms, for example H_2, N_2,..., etc., special considerations are needed in order to obtain the correct selection of quantum states. Consider first the simplest case H_2. The H nucleus, or proton, is an elementary particle and two protons are identical in the same sense as two electrons. From numerous experiments it is known that the proton also has a spin $s = \tfrac{1}{2}$ and obeys the Pauli exclusion principle. We can therefore conclude that the wave function of two protons must be antisymmetrical in the two particles if the spin wave function is included. The latter is exactly the same as for two electrons (Table 5, p. 84):

antisymmetrical: $\alpha(1)\beta(2) - \alpha(2)\beta(1)$, $s = 0$, ($\uparrow\downarrow$)

symmetrical: $\begin{cases} \alpha(1)\alpha(2) \\ \alpha(1)\beta(2) + \alpha(2)\beta(1) \\ \beta(1)\beta(2) \end{cases}$, $s = 1$, ($\uparrow\uparrow$)   (16)

s is the total spin of both protons.

The energy function for the motion of the nuclei is given by (7) and (8),

$$H = H_{\text{rot}} + H_{\text{vib}},$$

and we have seen that both parts have a sharp value (provided that the distance R is regarded as constant in H_{rot}). H_{vib} depends only on the absolute value $|\mathbf{R}| = R$ of the distance and involves a differentiation with respect to R only. In H_{rot} we have regarded R as constant. H_{rot} therefore only depends on the direction of $\mathbf{R}$ and involves differentiations with respect to the angles, θ, ϕ, say, formed by the vector $\mathbf{R}$ (with respect to a chosen frame of reference). The wave function consists then of a product

$$\psi = \psi_{\text{rot}}(\theta, \phi)\psi_{\text{vib}}(R) \qquad (17)$$

and

$$H_{\text{rot}}\psi_{\text{rot}} = E_{\text{rot}}\psi_{\text{rot}}, \qquad H_{\text{vib}}\psi_{\text{vib}} = E_{\text{vib}}\psi_{\text{vib}}.$$

The wave equation

$$H\psi = (H_{\text{rot}} + H_{\text{vib}})\psi_{\text{rot}}\psi_{\text{vib}}$$
$$= E_{\text{rot}}\psi_{\text{rot}}\psi_{\text{vib}} + E_{\text{vib}}\psi_{\text{rot}}\psi_{\text{vib}} = E\psi_{\text{rot}}\psi_{\text{vib}}$$

is then indeed satisfied for $E = E_{\text{rot}} + E_{\text{vib}}$.† $R = |\mathbf{r}_1 - \mathbf{r}_2|$

† Whenever the energy function is the sum of two parts, each of which depends on different variables, the wave function is a product of two functions depending on the two sets of variables separately. The case of two electrons without interaction is another example (Chapter V, section 2).

DIATOMIC MOLECULES

is evidently symmetrical in the two particles and therefore $\psi_{\text{vib}}(R)$ is symmetrical. Our symmetry considerations are concerned with ψ_{rot} only.

For an electron moving in a central field of force we have obtained the wave functions, as far as they depend on the direction of the radius vector **r**, in Chapters III and IV. When the angular momentum l is zero, $\psi = \psi(r)$ is independent of the direction. When $l = 1$ we have three wave functions corresponding to $m_z = -1$, 0, $+1$, namely $\psi_{\pm 1} = (x \pm iy)f(r)$, $\psi_0 = zf(r)$.

For a rotating molecule all we have to do is to replace the position vector **r**, by the vectorial distance of the two nuclei **R**. If we denote the Cartesian components of **R** again by x, y, z, and furthermore take into account that $R = \text{const.}$, we obtain

$$\psi_{\text{rot}} = \text{const.}, \quad \text{for } l = 0, \tag{18a}$$

$$\psi_{\text{rot}} = \begin{Bmatrix} \dfrac{x+iy}{R} \\ z/R \\ \dfrac{x-iy}{R} \end{Bmatrix} \times \text{const.}, \quad \text{for } l = 1. \tag{18b}$$

(x/R, etc., depend on the direction of **R** only, not on R itself.) If we interchange the two nuclei, $\mathbf{R} = \mathbf{r}_1 - \mathbf{r}_2$ changes sign. Hence x, y, z also change sign whereas R remains the same. We therefore see that

$$\psi_{\text{rot}} = \begin{cases} \text{symmetrical} & \text{for } l = 0 \\ \text{antisymmetrical} & \text{for } l = 1. \end{cases}$$

For $l > 1$ the symmetry properties are similar. We had mentioned on p. 42 that for d states ($l = 2$) the wave functions behave like x^2/R^2, xy/R^2, etc. They are evidently again symmetrical. In the general case the symmetry is as follows

$$\psi_{\rm rot} = \begin{cases} \text{symmetrical,} & l = \text{even} \\ \text{antisymmetrical,} & l = \text{odd.} \end{cases} \quad (19)$$

When we combine (19) with (16) and take the Pauli principle into account, we find that the rotational levels are connected with the total spin of the nuclei,

$$l \text{ even:} \quad s = 0, \quad (20)$$
$$l \text{ odd:} \quad s = 1.$$

Levels with l even and $s = 1$ (and also with l odd, $s = 0$) would have a symmetrical total wave function and are forbidden. The rotational levels with even l have only the $(2l+1)$-fold orbital degeneracy, whereas the levels with odd l have in addition a three-fold degeneracy due to the spin.

The spin directions of nuclei are usually hard to change. A large number of collisions (or adsorption at a solid surface) is required before a change of spin direction takes place. The two types of states behave almost as if they belonged to two different kinds of molecules. They are called *para-hydrogen* ($s = 0$) and *ortho-hydrogen* ($s = 1$). Para-hydrogen has only rotational levels with even l, ortho-hydrogen those with odd l. In ordinary hydrogen the two kinds of molecules occur in the ratio 1:3 because ortho-hydrogen has three times as many spin levels (on the average the l-degeneracy is the same).

Finally we consider briefly molecules consisting of higher atoms. A nucleus is known to consist of a number of protons and neutrons. The neutron has the same weight as a proton and has also spin $\frac{1}{2}$ and obeys the Pauli principle. The number of protons plus neutrons is the atomic weight A. Also two such composite nuclei are 'identical particles'. When we interchange two like nuclei with

DIATOMIC MOLECULES 121

atomic weight A, we interchange in reality A pairs of particles. Since for each interchange of either two protons or two neutrons the wave function changes sign, we have a change of sign A times. Thus, the wave function of the two nuclei will be symmetrical/antisymmetrical in the two nuclei if A is even/odd. We see that for even A the wave function of the two particles is *symmetrical*, in contrast to the cases considered so far. The spin of a nucleus is composed of the spins of the protons and neutrons, by the vector addition rule. When A is even, the total spin of the nucleus can only be $s = 0, 1, 2,...$, etc., whereas if A is odd the possible values are $s = \frac{1}{2}, \frac{3}{2},...$, etc. The actual value cannot be foretold but can be taken from experiment. We consider two examples: (i) oxygen has $A = 16$, and the spin of the nucleus is $s = 0$; (ii) nitrogen has $A = 14$ and $s = 1$. In both cases the total wave function of the two nuclei must be symmetrical. Now the total nuclear spin of O_2 is zero. Thus the orbital wave function of O_2 must be symmetrical. Hence we find

$$O_2: \text{ only even } l \text{ allowed.} \tag{21}$$

Half of the rotational levels are missing altogether. For nitrogen the total nuclear spin is obtained by combining the two spins $s = 1$. The total spin can be $s = 2, 1, 0$ (compare the addition rules in Chapter IV, sections 5 and 6). Without actually proving the fact the reader will readily believe (from what has been shown so far) that the spin wave function of the two nuclei is symmetrical when the total spin is $s = 2$ or 0, and antisymmetrical when $s = 1$. Thus the rotational levels are connected with the total nuclear spin as follows:

$$N_2: \quad \begin{array}{ll} l \text{ even}, & s = 0 \text{ or } 2 \quad \text{(ortho-nitrogen)} \\ l \text{ odd}, & s = 1 \quad\quad\quad \text{(para-nitrogen)}. \end{array} \tag{22}$$

The spin degeneracy is $2s+1$. Thus the levels $l =$ even are $5+1 = 6$-fold degenerate (apart from the l-degeneracy), the levels $l =$ odd are three-fold. Ortho- and para-nitrogen occur therefore in the ratio $2:1$.

The experimental facts underlying the above analysis of the rotational levels of nitrogen have played an important historical role in establishing the fact that nuclei are composed of protons and neutrons (and not of protons and electrons, as was believed before) and that the neutron has spin $\frac{1}{2}$ and follows the Pauli principle.

IX

THEORY OF HOMOPOLAR CHEMICAL BOND

1. The hydrogen molecule

ONE of the great puzzles confronting physicists and chemists in the days before wave mechanics was found, was the problem of the chemical bond. Roughly speaking, there are three types of chemical forces which lead to an attraction between atoms and molecules (although the distinction of these three cases is not always quite sharp), namely:†

(i) the heteropolar or ionic bond;
(ii) the homopolar or covalent bond;
(iii) the van der Waals forces.

As to (i) there is no difficulty in understanding why two ions with opposite charges, for instance Na$^+$ and Cl$^-$, attract each other. The van der Waals forces are usually very weak compared with the true chemical forces; they do not so much lead to the formation of molecules, but are rather responsible for keeping the molecules of a liquid together. We shall not consider them here.

The real problem lies in (ii). On the basis of classical physics it is quite impossible to understand why two neutral atoms, for instance two H atoms, form a molecule. Moreover, the covalent bond has the striking feature of showing *saturation*, expressed in the concept of *valency*.

† The so-called metallic bond is more complicated than the covalent bond and leads to the formation of crystals rather than molecules. It is also explicable by wave mechanics.

124 THEORY OF HOMOPOLAR CHEMICAL BOND

A H atom can bind *one* other H atom, but not two or three. A C atom can bind four H atoms but no more. Even if in classical physics sufficiently strong attractive forces between neutral particles were known, it would be quite impossible to understand why a third atom should not also be attracted by the two atoms already bound. The feature of saturation is entirely foreign to classical physics.

The problem has been solved on the basis of wave mechanics (Heitler and London, 1927). We shall see that wave mechanics not only explains the attraction between neutral atoms but also leads to a full understanding of the saturation properties.

We start with the simplest case, the H molecule. The most natural way to come to an understanding of the attraction between two H atoms will be the following: We consider the two atoms with their nuclei placed at two points a and b, say, at a distance R apart. Let the two atoms be both in their ground states. The total energy of the system will then differ from the sum of the internal energies of the two atoms (i.e. twice the energy of the ground state of H) by a certain *interaction energy*, $E(R)$, which will depend on the distance R. Only if R is very large will this interaction energy be zero. If now the atoms are really attracting each other $E(R)$ must *decrease* from the value 0 when $R = \infty$ to negative values when R is finite. At small distances R it is to be expected that $E(R)$ increases again and becomes positive, for otherwise no equilibrium distance would exist. Thus $E(R)$ will be a curve of the type shown in Fig. 30 or 31.

We can calculate $E(R)$ most conveniently by using the perturbation theory of Chapter VI. In fact the problem is very similar to the calculation of the energy-levels of the He atom, for, as in the He case, we have to do with *two*

THEORY OF HOMOPOLAR CHEMICAL BOND 125

electrons. The formulae for $E(R)$ will be very similar to the formulae for the perturbation energy of the levels of the He atom.

According to the general principles of the perturbation theory we have first to find the unperturbed wave function of the system. The electron of the first H atom whose

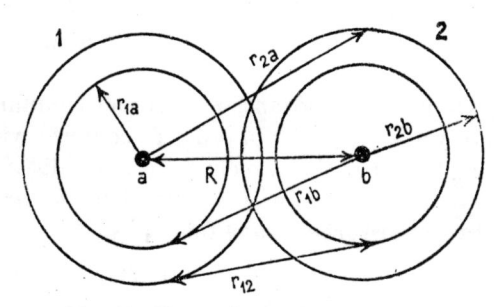

FIG. 32. Charge clouds of two H atoms.

nucleus is at a, say, is described by the wave function $\psi_a(1)$ given by equation (3) of Chapter III, p. 33, namely $\psi_a(1) = e^{-ar_{1a}}$ apart from the normalizing factor.† r_{1a} is here the distance of the electron from the nucleus which is placed at a. We have labelled this electron as number 1. Similarly, if we call the electron of the second H atom (whose nucleus is at b) number 2, then the electron of the second atom is described by the wave function $\psi_b(2)$. ψ_b is the same function as ψ_a, only it depends on the distance r_{2b} of the second electron from the nucleus at b. The charge cloud of the second electron is just shifted by the distance R to the point b as its centre. Fig. 32 shows the two charge clouds schematically. Except when R is very large the two charge clouds partly *overlap*. For the wave function of the

† The a occurring in the exponential is the reciprocal of the Bohr radius and has, of course, nothing to do with the a labelling the nucleus of the first H atom.

126 THEORY OF HOMOPOLAR CHEMICAL BOND

whole system we have then to take, as before, the product $\psi_a(1)\psi_b(2)$.

The crucial point is now that, just as in the case of the He atom, we obtain a second wave function by interchanging the roles of the two electrons. The second wave function is $\psi_a(2)\psi_b(1)$; it describes exactly the same system, namely two H atoms at a distance R, only the electron at a is labelled number 2 and the electron at b number 1. We have seen in Chapter V that in this case we are *forced* to take the symmetrical and antisymmetrical combinations of these two wave functions in order to conform with the fundamental fact that the two electrons are indistinguishable and cannot be affixed with names. Thus we have the following two undisturbed wave functions:

$$\psi_+ = \psi_a(1)\psi_b(2)+\psi_a(2)\psi_b(1), \qquad (1a)$$

$$\psi_- = \psi_a(1)\psi_b(2)-\psi_a(2)\psi_b(1). \qquad (1b)$$

In order to find the perturbation energy $E(R)$ we have to take the square of our wave functions (1), multiply it by the interaction function V, and integrate over all the coordinates of the two electrons. Now what is this interaction function V? In the He case V was simply the Coulomb interaction of the two electrons e^2/r_{12}. In our case the unperturbed potential energy is, however, only the interaction between the electron 1 and the nucleus a and between the electron 2 and the nucleus b, provided the electron 1 is attached to a and 2 to b. But, clearly, the electron 2 also interacts with the nucleus a and also 1 with b, and finally the two nuclei also interact with each other. All these interactions are part of the perturbation between the two atoms. Thus V in our case is

$$V = +\frac{e^2}{R}+\frac{e^2}{r_{12}}-\frac{e^2}{r_{1b}}-\frac{e^2}{r_{2a}}. \qquad (2)$$

Now (2) gives the Coulomb interaction between the two atoms provided that the electron at the nucleus a is labelled number 1. The wave function (1), however, contains also parts where the electron number 2 is at a and number 1 at b. Clearly in this case the interaction between the two atoms is not (2) (e^2/r_{2a} is then the internal potential energy of the atom a), but an expression obtained from (2) by interchanging 1 and 2:

$$V' = +\frac{e^2}{R} + \frac{e^2}{r_{12}} - \frac{e^2}{r_{2b}} - \frac{e^2}{r_{1a}}. \tag{2'}$$

Thus V is to be used as interaction for the first part of the wave function $\psi_a(1)\psi_b(2)$ and V' for $\psi_a(2)\psi_b(1)$.

The perturbation energy of the two atoms is now given by Chapter VI, equation (2), p. 88,

$$E(R) = \frac{\int V\psi^2\, d\tau}{\int \psi^2\, d\tau}, \tag{3}$$

where the integration is over the coordinates of both electrons and the 'V' to be inserted is either V equation (2) or V' equation (2') for the two parts of the wave function (1) respectively. We write out the numerator explicitly:

$$\int V\psi^2\, d\tau = \int V\psi_a^2(1)\psi_b^2(2)\, d\tau_1 d\tau_2 + \int V'\psi_a^2(2)\psi_b^2(1)\, d\tau_1 d\tau_2 \pm$$
$$\pm 2\int V\psi_a(1)\psi_b(1)\psi_a(2)\psi_b(2)\, d\tau_1 d\tau_2. \tag{4}$$

In the last term of (4) it may be doubtful whether to insert V or V' because here the product of the two parts of the wave function occurs and the electron 1 is 'partly' at the nucleus a and partly at b. This, however, makes no difference, for the last integral of (4) does not change if we insert V' instead of V. To see this we only remark that a change of notation of the variables over which we integrate does not change the value of the integral. If we then interchange the notation of the two electrons, $1 \to 2$ and $2 \to 1$, the product $\psi_a(1)\psi_a(2)\psi_b(1)\psi_b(2)$ does not change and V goes over into V', and therefore the integral does not change if we replace V by V'. In the same way we see that the first two integrals of (4) are equal.

Thus, using the same abbreviations as in Chapter VI,

$$C = \int \left(+\frac{e^2}{R} + \frac{e^2}{r_{12}} - \frac{e^2}{r_{1b}} - \frac{e^2}{r_{2a}} \right) \psi_a^2(1) \psi_b^2(2) \, d\tau_1 d\tau_2, \qquad (5)$$

$$A = \int \left(+\frac{e^2}{R} + \frac{e^2}{r_{12}} - \frac{e^2}{r_{1b}} - \frac{e^2}{r_{2a}} \right) \psi_a(1) \psi_b(1) \psi_a(2) \psi_b(2) \, d\tau_1 d\tau_2, \qquad (6)$$

the numerator of (3) is $2C \pm 2A$. We may assume that the wave functions of the two H atoms are normalized $\int \psi_a^2(1) \, d\tau_1 = \int \psi_b^2(2) \, d\tau_2 = 1$. The denominator of (3) then becomes $2 \pm 2S$, where

$$S = \int \psi_a(1) \psi_b(1) \psi_a(2) \psi_b(2) \, d\tau_1 d\tau_2. \qquad (7)$$

Unlike the case of the He atom, S does not vanish here. The orthogonality theorem (Chapter VI, section 3) which says that $\int \psi_a(1) \psi_b(1) \, d\tau_1 = 0$ applies only to the case where ψ_a and ψ_b are two different wave functions of the *same* atom belonging to two different stationary states, but in our case ψ_a and ψ_b are wave functions of two *different* atoms put at a distance R apart. Two such wave functions are not orthogonal. Looking at the charge clouds as shown in Fig. 32 we also see that the integral of the product of the two wave functions does not vanish unless the two atoms are so far apart that the two charge clouds do not overlap at all. (In this case also $E(R)$ vanishes.) On the

THEORY OF HOMOPOLAR CHEMICAL BOND

other hand, S is, except for very small distances, much smaller than 1. It would be equal to 1 only if $R = 0$ and therefore $\psi_a = \psi_b$, thus

$$S = \int \psi_a^2(1)\, d\tau_1 \int \psi_b^2(2)\, d\tau_2 = 1 \qquad (R = 0).$$

The 'overlap integral' S therefore only plays the part of a more or less minor correction (the denominator does not differ much from unity), and we can ignore it for a qualitative discussion.

We finally obtain for our interaction energy of the two H atoms the two formulae

$$E_+(R) = \frac{C+A}{1+S}, \qquad (8a)$$

$$E_-(R) = \frac{C-A}{1-S}, \qquad (8b)$$

where C, A, S are the integrals given by (5)–(7). All these integrals depend, of course, on R.

We see that, as in the case of the He atom, we obtain two different interaction energies. We know from Chapter V that the wave function of two electrons is always symmetrical in the coordinates of the two electrons if the two electrons have antiparallel spins, and it is antisymmetrical if the two electrons have parallel spins (see Table 5, p. 84). Now the interaction $E_+(R)$ belongs to the symmetrical combination (1a), whilst $E_-(R)$ holds for the antisymmetrical wave function (1b). Thus, if the two electrons have antiparallel spins ($^1\Sigma$ state) their interaction energy is $E_+(R)$. If the spins are parallel ($^3\Sigma$ state) the interaction energy is $E_-(R)$.† Again, as in the case of the He atom, we obtain two widely different interactions, their

† Since both H atoms are in S states it is evident that only Σ states can be formed for the H_2 system (see Chapter VIII, section 1).

difference is roughly $2A$. The reason for this difference is again the exchange phenomenon.

We now discuss our result: The integral C (5) can easily be interpreted. $\psi_a^2(1)$ is the charge density surrounding the nucleus a and $\psi_b^2(2)$ that surrounding the nucleus b. C therefore represents the Coulomb interaction of the charge cloud round a with the nucleus b (term e^2/r_{1b}), that of the charge cloud round b with the nucleus a, the interaction between the two charge clouds (e^2/r_{12}), and finally the interaction between the two nuclei. If we are at all to think of a classical picture that might give us an idea of the interaction of two neutral atoms, we should think that C would be the interaction energy. But we shall see that not C but A is the chief contribution and that the formation of a molecule rests essentially on A.

A does not permit such a simple interpretation as C. As we have seen, it is due to the fact that the two electrons are indistinguishable and can be exchanged. However, if we wish, we can, as in Chapter VI, section 2, define an *exchange charge density* $\rho = \psi_a(1)\psi_b(1)$. The term e^2/r_{12} of A represents, then, the Coulomb interaction of this exchange charge density with itself $\left(e^2 \int \dfrac{\rho(1)\rho(2)}{r_{12}} d\tau_1 d\tau_2\right)$, and the other terms may be interpreted as the interaction of the exchange charge with the nuclei, multiplied, though, by the total value of this exchange charge;

$$e^2 \int \rho(2) \, d\tau_2 \int \frac{\rho(1)}{r_{1b}} d\tau_1$$

is the contribution from the term e^2/r_{1b}.

In Fig. 33 the distribution of the exchange charge density is shown schematically.

In order to decide now whether the two H atoms attract or repel each other we have to evaluate the integrals C,

THEORY OF HOMOPOLAR CHEMICAL BOND 131

A, and S (the latter is of minor importance), and in particular to decide what their signs are. This is not quite so easy as in the case of the He atom. In C as well as in A both negative and positive contributions occur. The sign of C and A depends on whether the attraction of the charge clouds by the nuclei outweighs the repulsion between the two nuclei and between the charge clouds themselves. The chief features can, however, easily be seen from a look at Figs. 32 and 33.

In the first place, consider very large distances R. The exchange charge density ρ is then zero, because ψ_a and ψ_b do not overlap and $\psi_a(1)$ is very small everywhere where $\psi_b(1)$ is large and vice versa. Also all r's are then very large. Consequently both C and A vanish, as must be expected.

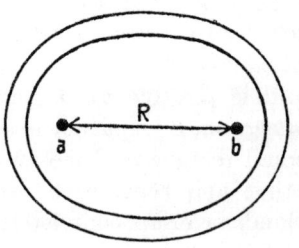

FIG. 33. Exchange charge.

Next consider medium distances, say of the order of magnitude of the Bohr radius $1/a$. At such distances the two charge clouds (Fig. 32) just overlap nicely because $1/a$ is also the distance from the nucleus at which the charge cloud is largest. First look at Fig. 33. Owing to the overlapping the exchange charge density is then quite large. We also see that certain parts of this exchange charge density are quite close to the nuclei and are therefore strongly attracted by them, giving rise to a large negative contribution to A. On the other hand, the average distance of the different parts of the exchange charge density is quite large (and so is the distance between the nuclei). Thus the positive contributions to A due to the repulsion of different parts of the exchange charge with each other (and also of the two nuclei) will be relatively smaller than the negative contributions. We thus conclude: *For medium distances A is negative.* Looking at Fig. 32 we see that for the same R the charge cloud 2, for example, is on the average farther away from the nucleus a than is the exchange charge from one of the nuclei at the nearest point, and also the two charge clouds are relatively far away from each other. We expect therefore that *C is numerically much smaller than A* for medium distances. The sign of $E(R)$ is therefore determined by A and is positive for E_- and negative for E_+. *E_+ means attraction, E_- repulsion of the two H atoms.* Finally, for very small distances R the two nuclei are very close to each other and repel each other strongly. The two charge clouds overlap completely, but are, of course, always a distance $1/a$ away from both nuclei. Far the largest contribution to C and A arises, then, from the Coulomb repulsion of the two nuclei. The factor e^2/R can be taken out of the integrals and these cancel in the numerator or denominator of (8). Thus for small R, $E_+(R) = e^2/R$, in both cases, the atoms repel each other strongly.

To sum up: Coming from large distances the *two H atoms attract (repel) each other if the spins of their electrons are antiparallel (parallel). For antiparallel spins the interaction energy $E_+(R)$ has a minimum at some distance of the order of magnitude of the Bohr radius.* For smaller distances the attraction goes over into repulsion.

These qualitative considerations are fully borne out by the calculations. The evaluation of the integrals C, A, S presents no difficulty but is rather cumbersome. We give the results in form of a graph. In Fig. 34 we have plotted the energies $E_+(R)$ and $E_-(R)$ as a function of the distance R using as units the Bohr radius $1/a$. As units for the energy we use electron volts (for comparison remember that the ionization energy of the H atom is 13·5 eV). C is also

THEORY OF HOMOPOLAR CHEMICAL BOND 133

plotted separately in Fig. 34. We see that $E_-(R)$ is always positive and therefore no stable molecule is formed if the two spins are parallel. $E_+(R)$ is indeed negative and far larger than C (the difference between E_+ and C is approximately A). $E_+(R)$ has a minimum at a distance $Ra = 1\cdot 6$

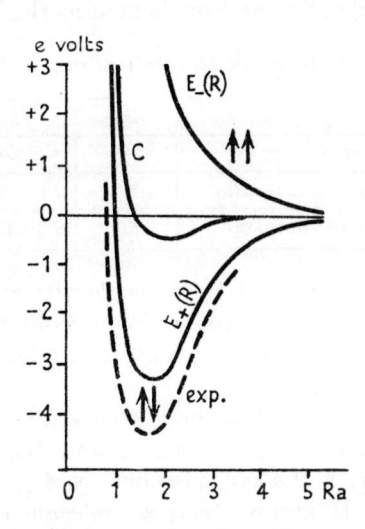

Fig. 34. Interaction of two H atoms.

and for smaller R, E_+ becomes positive. If, therefore, the two spins are antiparallel a *stable combination of the two H atoms is formed*, and this means the formation of a H_2 molecule. The minimum of $E_+(R)$ gives the *equilibrium distance* and the *binding energy* (or dissociation energy) of the H molecule. For comparison we have also plotted (dotted curve) the interaction curve of two H atoms as it is known from various experiments. We see that the theoretical curve fits the experimental one rather well. Table 9 gives the numerical data for the equilibrium distance, the binding energy, and the frequency of vibration ω near the minimum. (See Chapter VIII, section 3.) The agreement is as good as can be expected. We must not forget, of course, that we have used a perturbation method which only gives approximate results. There are more powerful and more complicated methods to calculate

TABLE 9. H_2 *molecule (theoretical and experimental)*

	Exp.	Theor.
Dissociation energy, eV	4·5	3·2
Equilibrium distance, Ra	1·4	1·6
Frequency of vibration	8·1	$9·0 \times 10^{14}$ sec.$^{-1}$

the binding energy of the H_2 molecule with great accuracy. But this was not our purpose. We set out to understand why two H atoms attract each other and form a molecule. This we have achieved:

The reason for the formation of a molecule is the quantum-mechanical exchange phenomenon, and the bulk of the binding energy is the exchange energy A.

When two H atoms meet, a molecule is not always formed. The two atoms repel each other if the spins of the two electrons are parallel. Indeed, if two H atoms collide, and if we do not know their relative spin directions, the chance is only 1:4 that they will attract each other and 3:4 that they repel each other. This is so because (Chapter V) there are three spin wave functions for the triplet state but only one for the singlet state. The existence of a repulsive interaction between two H atoms has also been found experimentally.

The H_2 molecule has, according to our theory, a spin zero. This also is in agreement with the facts, for H_2 is known to be diamagnetic and to have no magnetic moment.

THEORY OF HOMOPOLAR CHEMICAL BOND

As we have seen, the covalent bond rests on the quantum-mechanical exchange phenomenon. In view of the importance of this effect and the difficulty it represents for an understanding, we discuss now the question of what precisely is the physical meaning of this exchange. We have emphasized in Chapter VI that it cannot be fully understood by the picture of interacting charge clouds (which otherwise is useful) because A occurs with two different signs. Above all, the exchange is a typical *quantum*-mechanical effect and all attempts at finding a 'classical analogue' must fail. Nevertheless, apart from the large contribution to the energy, there is another aspect of the exchange phenomenon that might be helpful for its understanding. In Chapter I we have seen that to each energy E a frequency E/h is attributed. We may ask what is the meaning of the frequency attached to the exchange energy, viz. A/h? Is there anything that vibrates with this frequency? At first one might be tempted to think that the two electrons are really exchanging places with a frequency A/h, since A is due to the exchange of electrons. But this is not correct. The very basis of the exchange phenomenon is the fact that the electrons are indistinguishable and therefore an exchange of electrons is unobservable in principle. A/h can, though, be interpreted as the exchange frequency of spin directions. We can only describe this qualitatively, without proof. We may imagine that by means of some experiment we have obtained a knowledge of the individual spin directions of the two atoms, and that we found that at a certain instant of time the spin of the electron at a was ↑ (i.e. $m_{sz} = +\frac{1}{2}$) and that at b ↓. Once these individual spin directions are determined, this does not mean that they will remain the same in course of time. They will in fact exchange. We may perform the same experiment a little later and find, with a certain probability, that the spin at a is $\downarrow$ and that at b $\uparrow$. The probability is largest after a time $A/4h$. The change is periodical and the frequency of exchange is just A/h. The frequency A/h is thus the frequency of exchange of spin directions.

2. The saturation properties of the chemical bond

In the preceding section we have seen that neutral atoms can attract each other to form a molecule. It remains now to show that these chemical forces have the saturation properties known from the chemical facts. In this section we show this in a qualitative way, a more quantitative proof will follow in Chapter XI.

We consider first a simple example: the interaction of a He atom with a H atom. He in its ground state has two electrons, both in the lowest level. According to the Pauli principle the two electrons must have antiparallel spins and are in a ^1S state. If now the H atom is brought near so that interaction is taking place, the exchange phenomenon will come into play. In contrast to the case of two H atoms, however, there can be only *one* mode of interaction. For the vector addition rule for the spins shows at once that there is only one value for the total spin of the system, namely $\frac{1}{2}$, giving a $^2\Sigma$ state for the He–H system. Is this mode of interaction, then, attraction or repulsion? We can decide the question in the following way. Let the spin of the H electron be, for example, $\uparrow$. At first one might think that this electron could be exchanged with both the electrons of He. But this is not the case. If the H electron could be exchanged with the electron of He which has a spin $\downarrow$, then a state of the He atom would arise where both electrons have the *same spin* directions, namely $\uparrow\uparrow$, but this is forbidden by the Pauli principle. It follows that the H electron can only be exchanged with the one electron of He which has the *same spin* direction ↑ (Fig. 35).

Now we have seen in the case of two H atoms that the exchange of two electrons with parallel spins leads to

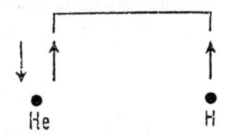

FIG. 35. Interaction of He and H.

repulsion. We conclude therefore: *A* He *atom and a* H *atom repel each other.* We expect that the interaction energy will be given by a formula of the type $E_-(R)$ (8b),

$$E(R) = \frac{C-A}{1-S};\qquad(9)$$

and the actual calculations show that this is the case, where, of course, C, A, S are not quite the same as in the case of two H atoms but are of similar type and have the same qualitative properties.

This consideration can at once be generalized to all rare gases. For the electron configuration of all rare gases consists of closed shells only, with all electrons arranged in pairs with antiparallel spin. Therefore a rare gas atom repels any other atom. This is in agreement with the chemical behaviour of the rare gases, which are known not to enter into chemical compounds with any other atom.

As a next step we consider the interaction of a H atom with a H_2 molecule which we assume to be already formed. As long as the H atom is at a distance much larger than the equilibrium distance of H_2 itself we may consider the H_2 molecule as a whole and ignore the fact that it consists of two H atoms. The electron state of H_2 is now, according to the preceding section, a singlet state with the two electrons forming a pair with antiparallel spin. It resembles therefore the electron configuration of He (at least as far as the spin is concerned). The electron of H can only be exchanged with the one electron of H_2 which has the same spin direction. It follows, then, that the H *atom is repelled by the* H_2 *molecule*.

This, then, is the explanation of the *saturation properties*. A H_2 molecule, once it is formed, is no longer capable of binding a third H atom, but any further H atom is repelled by it.

This consideration applies only as long as the H atom is comparatively far away from the H_2 molecule. If the distance between H and H_2 is comparable with the molecular distance of H_2 itself, the situation is more complicated and will be considered in Chapter XI, section 2, where the above conclusions will be verified by explicit calculation.

The repulsion of closed shells is also of great importance for the ionic bond. The ions Na^+ and Cl^- both have the same electron configuration as a rare gas; the only difference is that they are not neutral. They attract each other at large distances on account of the Coulomb forces which have a very large range, much larger than the exchange forces. At close distances they repel each other on account of the exchange forces of the closed shells. What is commonly called the 'size' of an ion is due to this repulsion. The exchange forces become strong very rapidly once the appropriate distance is reached.

X

VALENCY

1. Spin valency

WE must now generalize our considerations of Chapter IX to give an account of the valency and chemical binding properties of the higher atoms. The first and most important step in this direction will be quite easy and straightforward.

As we have seen in the two preceding sections, a molecule is only formed by two atoms if each atom can contribute an electron with a *free spin* so that a pair with antiparallel spins can be formed between the two atoms. If all the electrons of one atom are arranged in pairs within the atom (rare gases) so that the total spin of the atom is zero, no other atom can be bound. Moreover, as we have seen from the case of three H atoms, an atom with one free electron is capable of binding one other atom with a free electron, but no more. We therefore conclude that the valency properties of an atom must depend on the number of electrons with free spins (i.e. not used up in pairs).

In what circumstances may we then expect that an atom A, say, could bind two H atoms? Obviously, if the atom A has two free electrons which *have* or *can have* parallel spin, i.e. a total spin 1. The two H atoms can then be bound if their electrons arrange their spins in the direction opposite to that of A. So just one pair between A and each H atom is formed. Generally, we expect that an atom with n free spins can bind n H atoms. Fig. 36 (a) illustrates this for $n = 3$, for which nitrogen is an example.

Much the same considerations apply if an atom with

valency n forms a molecule with another atom with valency m. Then n or m pairs can be formed between the two atoms according to whether n or m is smaller. A saturated molecule is formed if $n = m$. Fig. 36 (b) shows this for the N_2 molecule with $n = 3$. Every *saturated molecule* is expected to have *spin* 0 and to be therefore diamagnetic. (For exceptions see the following section.)

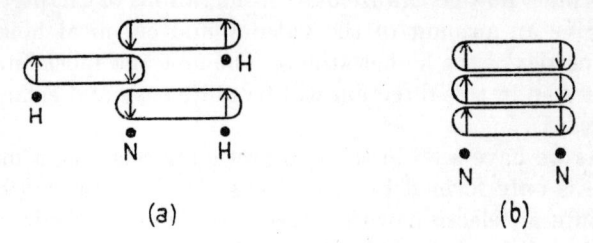

FIG. 36. Spin valency: NH_3 and N_2.

We thus see: What in chemistry is called the *valency of an atom is the number of free spins* of the atom or the total spin multiplied by 2. More precisely, this number is what is called the valency against hydrogen. What is called a bond between two atoms is a pair of electrons with antiparallel spin. A double bond means two pairs formed, etc.

The term 'have or can have parallel spin' requires a little more precision. If an atom has several electrons in the outermost shell, various atomic states arise from this configuration. For example, nitrogen has three p-electrons, giving rise to the states ^4S, ^2D, and ^2P. We can then define (a) either a valency of each specific atomic state, namely the number of electrons n with parallel spin. This is directly connected with the total spin of this state ($s = n/2$) or the multiplicity (($n+1$)-tuplet). Or else, (b), we can consider the whole configuration and define the valency of the configuration as the maximum number of electrons

VALENCY 141

which can arrange their spins in parallel.† In Chapter VII, section 2, we have seen that the lowest state of each configuration is usually that where the spins are, as far as possible, parallel. Thus the valency of the ground state (defined by (a)) will be identical with the valency of the lowest configuration (defined by (b)). The states ^4S, ^2D, ^2P, of nitrogen have valencies 3, 1, 1 respectively, and the valency of the p^3 configuration is 3.

We must still discuss to what extent excited states (or excited configurations) can be essential for the formation of molecules even for their ground state. Confining ourselves, to start with, to the lowest configuration, we can read off the valency of an atom directly from the spin of the ground state. In Table 7, p. 104, we have given the ground states for the atoms of the first period of the periodic table. In Table 10 we give the valency of these elements, as derived from the ground states by the above rule. n should then be the maximum number of H atoms that can be bound. In the last row we give the molecules of the type AH_n with the maximum number of H atoms as they are known to exist in chemistry.

TABLE 10. *Valency of elements of the first period*

	H	He	Li	Be	B	C	N	O	F	Ne
Total spin	1/2	0	1/2	0	1/2	1	3/2	1	1/2	0
Valency n from ground state	1	0	1	0	1	2	3	2	1	0
Exp. AH_n	H_2	..	LiH	[BeH]	BH	[CH_4]	NH_3	OH_2	FH	..

We see that the observed valencies agree in general with the expected ones, with two exceptions (indicated

† We shall see below (section 3) that for the treatment of the interaction of such atoms also two alternative starting points are possible, corresponding to the two views (a), (b).

by square brackets), namely Be and, most important, C. Carbon is very well known to have valency 4 (the whole of organic chemistry rests on this fact), whereas its ground state is ^3P with only two free electrons. In order to explain a valency 4 we should expect a state for the C atom with four free spins, i.e. a quintet state. Now we have seen in Chapter VII, section 2, that such a state really exists. It is a ^5S state. But this quintet state is not the ground state of C, it arises from the *excited electron configuration sp^3*. It will be seen in the following section that this excited state is really responsible for the four valencies of C.

We are thus led to consider not only the ground states of the atoms as responsible for their chemical properties but also their excited states. This will be of particular importance if an excited state has a higher valency than the ground state. In the following sections we investigate the influence of the excited states, and the case of carbon will be treated in detail as an example. (The case of Be is very similar.)

In addition to the free electrons an atom has also in general closed shells underneath. These would always repel the other atoms. Since they are much closer to the nucleus their influence will only be noticeable at small distances. At the equilibrium distance they will at most be able to diminish the binding energy of the molecule somewhat but not hinder the formation of a molecule.

Long before wave mechanics was known Lewis put forward a semi-empirical theory according to which the covalent bond between atoms was effected by the formation of pairs of electrons shared by each pair of atoms. We see now that wave mechanics affords a full justification of this picture, and, moreover, gives a precise meaning to these Lewis's electron pairs: they are pairs of electrons with antiparallel spin.

2. Crossing of atomic interaction curves. Valency of carbon

The valency 4 of carbon can only be understood as due to the excited ^5S state. One might think that it would be necessary first to spend the energy of excitation (about 4·2 eV, according to Table 8, p. 105) before the four H

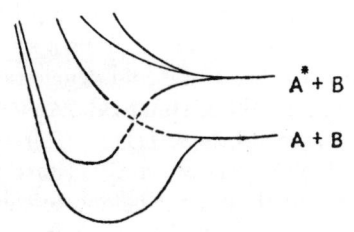

FIG. 37. Atomic interaction curves.

atoms can be bound. But this will be seen not to be the case. In order to understand the valency of carbon properly we investigate in this section the influence of the excited states quite generally.

Consider two atoms A and B both in specified atomic states. When these atoms are brought into contact they will interact in various ways, according to which of the molecular electronic states is formed. The variety of states which can be formed has been derived in Chapter VIII, section 1. Suppose we calculate the interaction energy by means of perturbation theory, much in the same way as we have done for two H atoms. Some of the modes of interaction will be attractive, some repulsive, according to how many electron pairs with antiparallel spin are formed.

It may now happen that an attraction curve starting from excited atomic states *crosses* a repulsion curve (or one with weak attraction) starting from the atoms in their ground states. In Fig. 37 such a case is indicated and the crossing region is dotted. In particular this will happen if, for instance, an atom in an excited state (A^*, say) has a larger valency than in its ground state and is able to form more pairs with its partner.

At the crossing-point a new *degeneracy* would arise. There would be two different states which at the particular distance have the same energy.

Now, in our previous exposition we have seen that whenever a degeneracy occurs there is a special reason for it. An atomic p state is three-fold degenerate because the electron moves in a field of spherical symmetry. The wave function in this case (Chapter III) is $xf(r)$, but on account of the spherical symmetry wave functions $yf(r)$ and $zf(r)$ must also exist and they must belong to the same energy. The same applies to the two-fold degeneracy of a molecular Π state, the reason being the axial symmetry of the molecular field. Whenever this symmetry is removed, for instance by an external magnetic field, the degeneracy is also removed and the state splits up into several states with different energies. In our case there is no such reason for the degeneracy at the crossing-point. We expect therefore that in reality *no such crossing-point can exist* (with the exception of the case mentioned below). In other words, the degeneracy in the would-be crossing-point is *de facto* removed and the two states split up.

If we are led to such a crossing by valency arguments of the kind used above or by calculating atomic interactions with the help of the perturbation theory of Chapter VI this only shows that in the neighbourhood of the crossing region this form of perturbation theory is inadequate. This latter only holds if the perturbation is small, in particular only if the state whose perturbation we wish to calculate is *far away* from all the other states. More precisely: the perturbation energy must be small compared

VALENCY

with the spacing of the atomic energy-levels. Only if the atomic interaction curves are far away from each other can we derive their qualitative behaviour from pure valency considerations. In a crossing region a new effect is taking place which, in some cases, modifies the results of section 1.

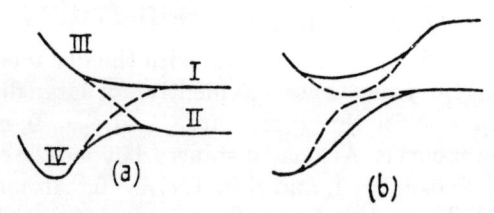

FIG. 38. 'Crossing' of atomic interaction curves.

We can describe qualitatively what must happen in the would-be crossing region thus: To avoid actual crossing the curves must give place to each other and deviate in such a way that they do not come very close. They virtually 'repel' each other. Beyond the crossing region their mutual influence diminishes again and the curves join the original branches. However, the first curve must then join what was originally part of the second curve and vice versa. This is shown in Fig. 38 (a): the dotted curves would cross, but the *real curves* as they *actually are* in nature, are the full-drawn ones.

A further point concerns the wave functions. Approximately, these are still products of atomic wave functions. Suppose the branches I and II of Fig. 38 (a) belong to wave functions $\psi_{A^*}\psi_B$ and $\psi_A\psi_B$ respectively, where A^* is an excited state of atom A, for example. If we ignore the mutual influence of the two curves, branch IV is a continuation of I, and III of II. In IV, therefore, the wave function is $\psi_{A^*}\psi_B$, not $\psi_A\psi_B$. The true curves start from I (or II), pass through an intermediate region but join the branches III (or IV) again. Thus $\psi_A\psi_B$ must go over *gradually* into $\psi_{A^*}\psi_B$, and $\psi_{A^*}\psi_B$ into $\psi_A\psi_B$. This means that in the intermediate region the wave functions must be linear combinations

$$c_{\mathrm{I}}\psi_{A^*}\psi_B + c_{\mathrm{II}}\psi_A\psi_B \quad \text{(for I–III)} \\ c'_{\mathrm{I}}\psi_{A^*}\psi_B + c'_{\mathrm{II}}\psi_A\psi_B \quad \text{(for II–IV)} \Bigg\}. \tag{1}$$

The coefficients c_{I}, c'_{I}, etc., change with the distance, as we pass along the true curves. Evidently, for large distances $c_{\mathrm{I}} = 1$, $c_{\mathrm{II}} = 0$ for the upper curve, and $c'_{\mathrm{I}} = 0$, $c'_{\mathrm{II}} = 1$, for the lower curve. At small distances, beyond the crossing region, $c_{\mathrm{I}} = 0$, $c_{\mathrm{II}} = 1$, and $c'_{\mathrm{I}} = 1$, $c'_{\mathrm{II}} = 0$. Although the portion IV is reached from atoms in their ground states (A and B) directly the wave function is here $\psi_{A^*}\psi_B$. This, of course, is only true if we have moved sufficiently far away from the crossing region.

All these conclusions can, of course, be verified by calculation; in fact a simple improvement of the perturbation theory of Chapter VI suffices to treat the above effect quantitatively, but this we need not carry out.

There is one exception to the above conclusions and this is if the two interaction curves in question belong to *different 'races'* (see p. 111). Then they can really cross. This can be seen in the following way. Supposing we start from the branch I of Fig. 38 (*a*). The spin and angular momentum are constant numbers which cannot change suddenly if we slowly change the distance of the atoms. Coming to the crossing region we meet a state with different spin or angular momentum. If we were to deviate on the full-drawn line to branch III, we would have to change either the spin or the angular momentum because branch III is the continuation of the curve II which belongs to a different race. Therefore the above effect can only happen if the two curves belong to the same race. Curves of different races are, in a way, foreign to each other; they may accidentally cross, then an 'accidental degeneracy' (not caused by symmetry reasons) occurs.

The above effect even takes place if the two curves would only come near to each other without actually crossing. Their distance is enlarged (Fig. 38 (b)). Thus, molecular attraction curves are further deepened (giving still stronger attraction) if, from some excited atomic state, a curve starts belonging to the same race. We now see the reason why we obtained too small a binding energy of the hydrogen molecule in Table 9, p. 134. This is due to the influence of the higher atomic states.

It is now easy to understand the four valencies of carbon. For simplicity we assume that the four H atoms have all the same distance R from C and are brought close to C simultaneously. The total energy of the system is then only a function of one distance R. The ground state of C is ^3P (s^2p^2 configuration). In this state the valency of C is 2 and the four hydrogen atoms cannot all be attracted. In fact, if they are all brought in contact with the C atom simultaneously they are attracted only very weakly (as a calculation shows). We consider in particular the state where the total spin is zero. Thus from the ground state of C an almost horizontal curve starts. Of all the excited states we need only consider the ^5S. The ^5S state has four valencies and is therefore able to attract all four H atoms strongly. This attraction curve, which naturally also has spin 0, is now bound to 'cross' the curve starting from ^3P. The situation is similar to Fig. 38 (a). Instead of crossing, the curve starting from the ground state is pushed down and leads to a stable molecule, whilst the curve actually starting from ^5S remains high.

Although it is the decisive influence of the excited ^5S state that makes it possible to bind four H atoms, the H atoms when removed from the carbon will leave the C atom in its ground state. And vice versa: When four H atoms collide with a C atom, either simultaneously (which in practice does not occur) or one after another, it is by no means necessary first to excite the C atom to the ^5S state. They can all be bound one after the other.

Similar to C is the N^+ ion. It has the same electron configuration as C and therefore four homopolar valencies. In addition it can bind a negative ion in a heteropolar bond. In this way molecules like $(N^+H_4)Cl^-$ can be formed.

Other examples for the mutual influence of such molecular interaction curves are the molecules C_2 and O_2. From section 1 we should expect that all saturated molecules are in singlet states. Whilst this is in general found to be the case, these two molecules are exceptions. C_2 is in a $^3\Pi$ and O_2 in a $^3\Sigma$ state. In fact, oxygen is well known to be paramagnetic, which means that it has a permanent magnetic dipole moment and must therefore have a spin different from zero. We consider briefly as an example the case of C_2. If both C atoms are in their ground states ^3P the strongest attraction curves should be the singlets, while the triplets with only one pair formed should lie higher but should also be slightly attractive. The low excited states of C are ^1D, ^1S, and ^5S. All these when interacting with the second C atom in ^3P give rise only to triplets (or higher multiplicities). In particular, from $^5S + {}^3P$ a triplet curve with strong attraction should arise because two pairs are formed. This triplet curve will cross, or at any rate come near to, the triplet with weaker attraction starting from $^3P + {}^3P$. The result is that the latter curve is strongly pushed down *even below* the singlet curve from $^3P + {}^3P$. This is shown qualitatively in Fig. 39 (only the essential curves are drawn; many more curves start in fact from the atomic states in question). A triplet then becomes the ground state of C_2. More detailed considerations are needed to show that it is actually a $^3\Pi$ (and not $^3\Sigma$ or $^3\Delta$) state.

Finally it should be mentioned that ionic states may also contribute to increase the binding energy, even in a homopolar molecule. Although the ionization energy is

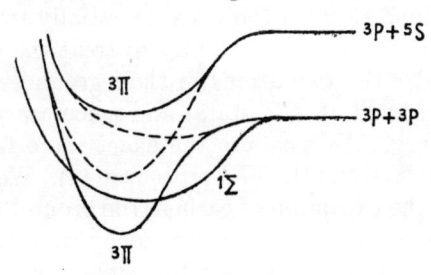

FIG. 39. Molecular states of C_2.

then usually high this is to some extent compensated by the attraction of the ions and the ionic curve may come down deep enough to strengthen the bond of the homopolar curve.

It is seen that the crossing effect discussed above modifies the conclusions drawn from pure valency considerations in some cases. It changes in special circumstances the order of the molecular terms and always contributes to *increase* the binding energies. But it never replaces a valency in the sense that it would make the formation of a molecule possible without proper electron-pair binding. In particular, the crossing effect makes it possible that a moderately low excited state with higher valency can play its full part without an excitation energy first being spent. In broad outline, therefore, the *chemical properties of an atom are determined by the maximum spin valency of the few lowest states of the atom*. In this way the results of the theory are in full agreement with the facts.

3. Interaction in diatomic molecules

It will be useful to corroborate the considerations of the preceding sections, especially of section 1, by explicit calculation. Before we do this we must first make some remarks about the approximations to be used. For atoms with more than one electron there are essentially two different approaches: The most direct way to treat the interaction is to consider the two atoms in their ground state (or in another specified atomic state) and calculate their interaction energy. We may use the exact wave functions of each atom. We call this the procedure (*a*). We have seen that, with the exception of carbon, the ground state of the atom has in general the correct valency, and we expect to understand the formation of the molecules observed in chemistry. However, in this way we neglect the influence of the excited states which (section 2) is negligible only when the curves starting from these states are sufficiently far away from the molecular curve considered. For large atomic distances where the interaction energy is still small, this is always the case.

On the other hand, the low excited states (again with the exception of carbon) all arise from the same electron configuration and their energy difference is due to the interaction of the electrons within the atom. If we neglect the latter (to start with) several atomic states will coincide and the electrons in the outermost shell will be free to arrange themselves and their spins (as far as the Pauli principle permits) in this shell. Let now an atom in such a configuration interact with other atoms. In the lowest molecular state the electrons will then arrange themselves so as to give an optimum binding rather than form specific atomic states. In this way we evidently take into account all the atomic states of the same configuration.

We call this the procedure (*b*). It is a good approximation if the atomic states lie very close so that the interaction curves starting from the ground state and the excited states are also close. The interaction of the electrons within each atom can be taken into account as a perturbation but the unperturbed wave functions of each atom are used (i.e. products of single electron wave functions).

A great advantage of (*b*) is that this method easily accounts for the directional properties of the valencies. These are, at least partly, due to the influence of the excited states. We shall treat them in Chapter XI, section 3.

The truth usually lies somewhere between (*a*) and (*b*). Better than both (*a*) and (*b*), but also more complicated, is the procedure described in section 2: consider the curves starting from specified atomic states and take their mutual influence as well as the correct excitation energy into account. But this we shall not carry through quantitatively.

For many atoms, for example N, (*a*) is a sufficiently good approximation if we do not insist on very accurate results and are mainly interested in the binding properties, although the excitation energies are of the same order of magnitude as the binding energies of the molecule. The reason is that the excited states have smaller valency and give rise to repulsion. Thus the attraction curve starting from the ground state and the repulsion curves starting from the excited states are far away from each other in the neighbourhood of the equilibrium distance.

Special considerations are needed for carbon. The excited ^5S state is all-important but the other states arising from the configuration sp^3 lie very high and (*b*) would be a poor approximation. For molecules which require only two (or one) of the four valencies of carbon (like CH, CH_2, and C_2) the co-operation of ^3P and ^5S is essential. On the other hand, for molecules in which three or four valencies are needed the procedure (*a*) (applied to ^5S), although it cannot be too accurate, may well be used, provided we are only interested in the stable molecule. We have seen in section 2 that beyond the region where the ^5S curve would cross the curves starting from ^3P, i.e. near the potential minimum of the molecule, the wave function of the carbon atom is close to that of the ^5S state. Thus the part of the interaction curve describing the molecule can be calculated by taking into account only the ^5S state, but it must be borne in mind that this is not permissible if we study the asymptotic behaviour for large distances.

Let us then use in this section method (*a*). Let *a*, *b* be two atoms with *n*, *m* valency electrons with parallel spin. The wave functions of *a* and *b* are then $\psi_a(1,...,n)$ and $\psi_b(n+1,...,n+m)$, where we have numbered the electrons of *a* 1 to *n*, those of *b* $n+1,..., n+m$. Since the spins in each atom are parallel their spin wave functions are entirely symmetric and thus ψ_a and ψ_b must be antisymmetric in all their electrons. We may assume $n \geqslant m$, and then between *a* and *b* a number of pairs, say *p* pairs, can be formed. *p* ranges from 0 to *m*. Accordingly, the total spin of both atoms is

$$s = \frac{n+m}{2} - p. \qquad (2)$$

The interaction energy will be expressed by certain integrals which are a generalization of the Coulomb and exchange integrals C and A of Chapter IX, section 1. The Coulomb integral is

$$C = \int \psi_a^2 \psi_b^2 V \, d\tau, \qquad (3)$$

where V is the interaction of all the electrons of *a* with those of *b* and the nucleus of *b*, etc. The integration is over the coordinates of all electrons ($d\tau \equiv d\tau_1 d\tau_2 ... d\tau_{n+m}$).

VALENCY

There is also an exchange integral

$$A = \int \psi_a(1,\ldots,n)\psi_b(n+1,\ldots,n+m) \times \\ \times V\psi_a(n+1,2,\ldots,n)\psi_b(1,n+2,\ldots,n+m)\,d\tau, \quad (4)$$

in which the electrons numbered 1 and $n+1$ are exchanged. The reader will easily convince himself that the integral has the same value if any other two electrons of a and b are exchanged.

In addition integrals occur in which, for example, two pairs of electrons are exchanged, $1 \leftrightarrow n+1$, $2 \leftrightarrow n+2$. In fact all kinds of permutations of the $n+m$ electrons occur. All these integrals are numerically much smaller than A, as a closer examination shows. We shall neglect them in the following. Also the 'overlap integral' S (p. 129) may be neglected.

Without very excessive labour it will not be possible to calculate the values of these integrals numerically. As, however, the structure of the integrals is similar to that for hydrogen, we are justified in assuming that their behaviour, as a function of the distance, is similar. Thus we can assume that for large and medium distances both C and A are negative and that A is numerically considerably larger than C. The formation of a molecule, and the strength of the bond, therefore depends on the sign and the factor with which A occurs.

How does the interaction energy then depend on C and A? This is most easily derived by the method explained in Chapter XI, sections 1 and 4. We anticipate the result ((5) is derived on p. 183):

$$E = C + A[p - (n-p)(m-p)]. \quad (5)$$

When the maximum number of pairs is formed, $p = m$, we have
$$E = C + pA. \quad (6)$$

The exchange integral occurs with a factor p corresponding to the fact that a p-fold bond between the two atoms exists. For smaller p the factor of A decreases and then becomes negative (repulsion). Consider, for example, two N atoms, $n = m = 3$. Both are in ^4S states and therefore only Σ states are formed. The energies are

$$E = \begin{matrix} C+3A, & p=3, & s=0, & ^1\Sigma, \\ C+A, & p=2, & s=1, & ^3\Sigma, \\ C-3A, & p=1, & s=2, & ^5\Sigma, \\ C-9A, & p=0, & s=3, & ^7\Sigma. \end{matrix} \qquad (7)$$

Apart from the lowest state with a triple-bond ($p = 3$) there is an excited, but still stable, triplet state ($p = 2$). Although two bonds still exist, the energy is only $C+A$. Excited triplet states of N_2 are indeed known. When only one bond is formed ($p = 1$) the energy is $C-3A$ which certainly means repulsion. This is rather surprising because one pair *is* formed, between the two atoms. It is a general feature of the theory that the binding energy between two atoms is very much diminished, or even overcompensated, if both atoms have either free unsaturated valencies or are bound jointly to a third atom. Our result is in agreement with the facts, though, because no molecular quintet state of N_2 is known.

XI

POLYATOMIC MOLECULES

1. Interaction of several atoms with one valency electron

To understand the chemical behaviour of a system of atoms in polyatomic molecules as well as their reactions, we calculate the interaction energy. This will depend on the distances between each pair of atoms. We shall see that it can be expressed in terms of Coulomb and exchange integrals.

To begin with, we consider a number of atoms $a, b,...$, etc., placed somehow in space and with one valency electron each. It is convenient to assume an even number of atoms. The case of an odd number can be reduced to this by assuming one further atom and putting the interaction integrals of this last atom with all other atoms equal to zero.

Atom a has then a wave function $\psi_a(1)\alpha(1)$ or $\psi_a(1)\beta(1)$ according to its spin direction. A typical total wave function is then

$$\psi_a(1)\psi_b(2)\psi_c(3)...\alpha(1)\beta(2)\alpha(3).... \qquad (1)$$

Since we are most interested in molecule formation, let the number of α-factors and β-factors be equal (total spin zero). Further wave functions, all degenerate with (1), are obtained by a different numbering of the electrons attributed to $a, b,...$, as well as by changing some of the α-factors into β and vice versa.

As a first step, we can now construct spin wave functions such that between specified pairs of atoms a bond, or an electron pair with anti-parallel spin, is established. For example, we can have a bond between a and b and also between c and d. If the electrons are numbered as in (1) this spin wave function is (omitting factors relating to the remaining atoms)

$$\phi_{\mathrm{I}} = \tfrac{1}{2}[\alpha(1)\beta(2)-\alpha(2)\beta(1)][\alpha(3)\beta(4)-\alpha(4)\beta(3)]. \quad (2a)$$

Alternatively, bonds can be established between a and c and between b and d, and finally we could have a–d and b–c bonds. The corresponding spin functions are

$$\phi_{\mathrm{II}} = \tfrac{1}{2}[\alpha(1)\beta(3)-\alpha(3)\beta(1)][\alpha(2)\beta(4)-\alpha(4)\beta(2)], \quad (2b)$$

$$\phi_{\mathrm{III}} = \tfrac{1}{2}[\alpha(1)\beta(4)-\alpha(4)\beta(1)][\alpha(2)\beta(3)-\alpha(3)\beta(2)]. \quad (2c)$$

However, these three spin functions are not independent. By multiplying out the brackets one sees that

$$\phi_{\mathrm{III}} = \phi_{\mathrm{II}}-\phi_{\mathrm{I}}. \quad (3)$$

Thus one of the three spin functions is a linear combination of the other two. The factors $\tfrac{1}{2}$ have been added in order that the ϕ's are normalized to unity

$$\int \phi_{\mathrm{I}}^2 = \int \phi_{\mathrm{II}}^2 = \int \phi_{\mathrm{III}}^2 = 1, \quad (4)$$

when use is made of $\int \alpha\beta = 0$, $\int \alpha^2 = \int \beta^2 = 1$ for each electron (see p. 96). It is important to realize, however, that ϕ_{I}, ϕ_{II}, ϕ_{III} are not orthogonal to each other. Multiplying (3) by ϕ_{I}, ϕ_{II}, ϕ_{III} in succession and using (4), we easily obtain

$$\int \phi_{\mathrm{I}}\phi_{\mathrm{II}} = \int \phi_{\mathrm{II}}\phi_{\mathrm{III}} = -\int \phi_{\mathrm{I}}\phi_{\mathrm{III}} = \tfrac{1}{2}. \quad (5)$$

We can represent the three spin functions graphically in a very simple manner, indicating already their connexion with chemistry. We connect the atoms between which a pair is formed by a dash. Since, however, the sign of the

POLYATOMIC MOLECULES 157

spin functions matters, we must also attribute an arrow to each dash. Then the three spin functions (2) and the relation (3) can be represented as follows:

$$\begin{matrix} a & b \\ & \times & \\ c & d \end{matrix} \;=\; \begin{matrix} a & b \\ \downarrow & \downarrow \\ c & d \end{matrix} \;-\; \begin{matrix} a \longrightarrow b \\ \\ c \longrightarrow d \end{matrix} \qquad (3')$$

$$\phi_{\mathrm{III}} \qquad \phi_{\mathrm{II}} \qquad \phi_{\mathrm{I}}$$

We can choose two of the three spin functions as an 'independent basis', for example ϕ_{I} and ϕ_{II}. We call a wave function with fixed bonds, like ϕ_{I}, ϕ_{II}, or ϕ_{III} a *valency structure*. If we have more than four atoms, more valency structures exist, and for those relating to any four atoms a linear relation (3) or (3') exists, but there are no further linear relations. A basis of independent structures is conveniently obtained if we imagine the atoms all arranged on a circle (on paper, this has nothing to do with the true arrangement in space). Then a structure where two valency dashes cross (as in ϕ_{III}, (3')) can be expressed by those where no dashes cross. An independent basis is obtained by constructing all possible valency structures in which no two dashes cross. Complete wave functions are now

$$\psi_{\mathrm{I}} = \psi_a(1)\psi_b(2)\ldots\phi_{\mathrm{I}}, \qquad \psi_{\mathrm{II}} = \psi_a(1)\psi_b(2)\ldots\phi_{\mathrm{II}}, \text{ etc. (6)}$$

These wave functions are not yet antisymmetric. This is easily remedied. We add to ψ_{I}, for example, all those wave functions arising by a permutation of the numbering of the electrons with the proper sign. If we change the numbering $1, 2 \to 2, 1$ of the two electrons located in (6) at atoms a, b we obtain $\psi_a(2)\psi_b(1)\ldots\phi_{\mathrm{I}}(2, 1, \ldots)$ and this has to be added to (6) with a minus sign. Similarly, the two electrons located in (6) at any two atoms can be exchanged and there is a corresponding term to be added with a minus sign. For brevity we write these terms $T_{ab}\psi_a\psi_b...T_{ab}\phi_I$, etc., where T_{ab} has the function of exchanging the numbering of the electrons at a and b. This exchange is to be carried out in the orbital functions $\psi_a\psi_b...$ as well as in the spin function ϕ_I. There are also more complicated permutations in which three or more electrons are involved. For example, the numbering 1, 2, 3 can be permuted into 2, 3, 1, or 1, 2, 3, 4 → 2, 1, 4, 3, etc. All these occur in the completely antisymmetric wave function with proper signs but we need not write them down explicitly. Thus from (6) we obtain an antisymmetric wave function, again denoted by ψ_I,

$$\psi_I = \psi_a(1)\psi_b(2)...\phi_I(1, 2...) - \sum_{a,b} T_{ab}\psi_a\psi_b...T_{ab}\phi_I + ... \quad (7)$$

The sum a, b is to be extended over all pairs of atoms. For each valency structure we have then one unperturbed antisymmetric wave function of the type (7). If N such independent valency structures exist, we have N degenerate wave functions and they all belong to a total spin $s = 0$, because all valencies are saturated.

Our next task is to calculate the interaction energy. For this purpose we must now apply a different method, for the following reason: From the N degenerate wave functions $\psi_I,...,\psi_N$, we can, of course, also form linear combinations like

$$\psi = c_I\psi_I + c_{II}\psi_{II} + ... + c_N\psi_N. \quad (8)$$

In previous cases, we could find the 'correct' linear combination by mere considerations of symmetry, but here no such argument presents itself. Yet, when the atoms interact, the degeneracy is surely removed, and for each energy state the wave function must be, approximately, a definite combination of the type (8), with fixed coefficients and we have no reason to think that these wave functions are just the original $\psi_\mathrm{I}, \psi_\mathrm{II},\ldots$. The coefficients $c_\mathrm{I},\ldots$ must be determined from the problem itself, and we shall see that they even depend on the arrangement of the atoms in space (their distances, etc.).

We proceed as follows: supposing ψ is the *exact* wave function of the whole system, H the complete energy function comprising the kinetic energy of all electrons and the whole potential energy, including the interaction between the atoms, then the wave equation is satisfied $E\psi = H\psi$ (see p. 27). Now the energy function H consists of the internal energy of the atoms H_0 say, and the perturbation V, which in our case is the interaction between the atoms, $H = H_0 + V$. Similarly E is the value of the internal energy of all atoms E_0, say, plus the perturbation energy ΔE which we wish to calculate. If we denote the unperturbed wave function (8) by ψ_0 then $E_0\psi_0 = H_0\psi_0$. Now the exact wave function ψ differs, of course, from ψ_0, but if we confine ourselves to an approximate determination of ΔE, in the sense of perturbation theory, we can identify ψ with ψ_0 (see Chapter VI). Then we can subtract E_0 from E and H_0 from H and we find that the perturbation energy ΔE and the interaction function must also satisfy a wave equation *approximately*

$$\Delta E \psi_0 = V \psi_0. \tag{9}$$

Below we write E again for ΔE. ψ_0 is an unperturbed wave function of type (8) with as yet unknown coefficients $c_\mathrm{I},\ldots$. For ψ_I, etc., (7) is to be substituted.

V is the interaction between each pair of atoms

$$V = V_{ab} + V_{ac} + V_{bc} + \ldots. \tag{10}$$

To make use of (9) we multiply this equation by $\psi_a(1)\psi_b(2)\ldots$ and integrate over the coordinates of all electrons. Since

$\int \psi_a^2(1) \, d\tau_1 = 1$, etc., we first obtain on the left a term equal to unity multiplied still by the spin function

$$c_\mathrm{I}\phi_\mathrm{I} + c_\mathrm{II}\phi_\mathrm{II} + \ldots .$$

The terms in which two electrons are exchanged give rise to integrals of the type

$$S_{ab} = \int \psi_a(1)\psi_b(2)\psi_a(2)\psi_b(1) \, d\tau_1 d\tau_2.$$

These are the 'overlap integrals' already encountered in Chapter IX. They are fairly small and do not play an important role. For simplicity we shall neglect them in the following. Thus the left-hand side of (9) reduces to

$$E(c_\mathrm{I}\phi_\mathrm{I} + \ldots + c_N\phi_N). \tag{11}$$

On the right-hand side we encounter similar integrals containing V. First the term without T_{ab} yields

$$\begin{aligned}
C &= \int (V_{ab} + V_{bc} + \ldots)\psi_a^2(1)\psi_b^2(2)\psi_c^2(3) \, d\tau_1 d\tau_2 d\tau_3 + \ldots \\
&= \int V_{ab}\,\psi_a^2(1)\psi_b^2(2) \, d\tau_1 d\tau_2 + \int V_{bc}\,\psi_b^2(2)\psi_c^2(3) \, d\tau_2 d\tau_3 + \ldots \\
&= C_{ab} + C_{bc} + \ldots, \tag{12}
\end{aligned}$$

where use is made of the fact that V_{ab} is independent of electron 3 and therefore $\int \psi_c^2(3) \, d\tau_3 = 1$. (12) is nothing but what we have called the Coulomb energy in Chapter IX. It represents the sum of the Coulomb interactions betweeen the charge clouds of the atoms. (12) occurs also with the factor $c_\mathrm{I}\phi_\mathrm{I} + \ldots + c_N\phi_N$.

Next consider the term containing T_{ab}. It is given by

$$\begin{aligned}
A_{ab} &= \int \psi_a(1)\psi_b(2)\psi_c(3)\ldots V \psi_a(2)\psi_b(1)\psi_c(3)\ldots d\tau \\
&= \int \psi_a(1)\psi_b(2)V_{ab}\,\psi_a(2)\psi_b(1) \, d\tau_1 d\tau_2 + \\
&\quad + \int \psi_a(1)\psi_b(1) \, d\tau_1 \int \psi_a(2)\psi_b(2)V_{bc}(23)\psi_c^2(3) \, d\tau_2 d\tau_3 + \ldots .
\end{aligned} \tag{13}$$

The first integral of (13) is exactly the exchange integral of Chapter IX, section 1. It refers to the interaction of a and b only and depends on the distance a–b. The second integral is of a new type. It can be described as the interaction of the charge cloud $\psi_c^2(3)$ of c with the 'exchange charge' $\psi_a(2)\psi_b(2)$, but this is multiplied by the overlap integral $\int \psi_a(1)\psi_b(1)\,d\tau_1$. The second term of (13) is therefore rather smaller than the first.† Thus we expect A_{ab} to be not very different from the exchange integral in Chapter IX. A_{ab} occurs on the right-hand side of (9) with the factor $-T_{ab}(c_\mathrm{I}\phi_\mathrm{I}+...+c_N\phi_N)$.

In addition further contributions arise from terms in which three or more electrons are interchanged. An example is

$$\int \psi_a(1)\psi_b(2)\psi_c(3)V\psi_a(2)\psi_b(3)\psi_c(1)\,d\tau. \tag{14}$$

Since each term of V depends only on two electrons, any such integral contains an overlap integral as factor and is again comparatively small. We shall neglect this type of integral also.‡

With the above simplifications we obtain from (9)

$$(E-C)(c_\mathrm{I}\phi_\mathrm{I}+...+c_N\phi_N)+\sum_{a,b}A_{ab}T_{ab}(c_\mathrm{I}\phi_\mathrm{I}+...+c_N\phi_N)=0. \tag{15}$$

Our next task will be to find out what is the effect of the exchange T_{ab} on the spin functions $\phi_\mathrm{I},...$. This will be very easy indeed and we shall see that $T_{ab}\phi_\mathrm{I}$ is itself a linear combination of the ϕ's. Thus (15) is an equation between the ϕ's. Since these spin functions are independent, the coefficients of each ϕ must vanish. (15) then splits up

† Since the interaction energy is expressed directly in terms of A_{ab}, etc., there is no need to neglect these terms, they may be included in A_{ab}.

‡ It is noteworthy that (14) depends on all the distances R_{ab}, R_{ac}, R_{bc} at the same time. Thus in the theory of the chemical bond there are not only forces between two atoms, but also three-body forces, etc.

into N linear equations for the unknown coefficients $c_\mathrm{I},\ldots,c_N$. From these both the c's and the energy E can be determined. If, however, we only want to know the energy we can put the problem even more simply. Instead of the coefficients $c_\mathrm{I},\ldots$ we regard the ϕ's themselves as unknowns. (15) is then evidently satisfied if we put, for each ϕ,

$$\left.\begin{array}{c}(E-C)\phi_\mathrm{I} + \sum_{a,b} A_{ab} T_{ab} \phi_\mathrm{I} = 0 \\ (E-C)\phi_\mathrm{II} + \sum_{a,b} A_{ab} T_{ab} \phi_\mathrm{II} = 0 \\ \cdots\cdots\cdots\cdots\cdots \\ (E-C)\phi_N + \sum_{a,b} A_{ab} T_{ab} \phi_N = 0\end{array}\right\}. \quad (16)$$

Since these are N linear homogeneous equations for the N unknowns ϕ the determinant of the coefficients of the ϕ's must vanish. This will provide an equation for E.

What is now the effect of the exchange T_{ab} on each ϕ? This can be read off from the valency structures (3′) immediately, without calculation. Each valency dash (with direction) represents a pair of electrons formed between specified atoms. All we have to do to obtain $T_{ab}\phi$ is to exchange the two end-points a, b of the two valency dashes starting from (or ending in) a and b, retaining the directions and keeping, of course, the other ends of these valency dashes (atoms $c, d,\ldots$) fixed. If, in ϕ, a valency dash connects a and b then T_{ab} just reverses its direction and $T_{ab}\phi = -\phi$. If ϕ is a structure such that a is connected with c and b with d, then $T_{ab}\phi$ is a structure where b is connected with c and a with d.

Consider as an example four atoms with the two spin functions ϕ_I, ϕ_II, (3). We have

$$T_{ab}\phi_\mathrm{I} = -\phi_\mathrm{I} = T_{cd}\phi_\mathrm{I}$$
$$T_{ab}\phi_\mathrm{II} = T_{ab}\downarrow\downarrow = \diagtimes = \phi_\mathrm{III} = \phi_\mathrm{II}-\phi_\mathrm{I} = T_{cd}\phi_\mathrm{II},$$
$$(17a)$$

POLYATOMIC MOLECULES 163

and in a similar way

$$\left.\begin{array}{l}T_{ac}\phi_\mathrm{I} = T_{bd}\phi_\mathrm{I} = -\phi_\mathrm{III} = \phi_\mathrm{I}-\phi_\mathrm{II}\\ T_{ac}\phi_\mathrm{II} = T_{bd}\phi_\mathrm{II} = -\phi_\mathrm{II}\\ T_{bc}\phi_\mathrm{I} = T_{ad}\phi_\mathrm{I} = \downarrow\downarrow = \phi_\mathrm{II}\\ T_{bc}\phi_\mathrm{II} = T_{ad}\phi_\mathrm{II} = \phi_\mathrm{I}\end{array}\right\}. \quad (17b)$$

We now insert (17) into (16). With the abbreviations

$$u = A_{ab}+A_{cd}, \quad v = A_{ac}+A_{bd}, \quad w = A_{bc}+A_{ad} \quad (18a)$$

we obtain

$$\left.\begin{array}{l}(E-C-u+v)\phi_\mathrm{I}+(w-v)\phi_\mathrm{II} = 0\\ (-u+w)\phi_\mathrm{I}+(E-C+u-v)\phi_\mathrm{II} = 0\end{array}\right\}. \quad (18b)$$

It follows that the determinant of the coefficients of the unknowns ϕ_I, ϕ_II must vanish:

$$\begin{vmatrix} E-C-u+v & w-v \\ -u+w & E-C+u-v \end{vmatrix} = 0, \quad (19)$$

or, explicitly,

$$(E-C)^2-(u-v)^2+(w-v)(u-w) = 0.$$

Hence we obtain two energy values

$$E = C \pm \sqrt{\{u^2+v^2+w^2-uv-uw-vw\}}. \quad (20)$$

The twofold degeneracy, due to the two possible valency structures, is removed: we have two different energies. (20) is the general formula for the interaction of four electrons. We shall use it in sections 2 and 3 for two different purposes.

It is of some interest to determine also the coefficients c_I, c_II in (8). Clearly only the ratio is determined. We give the result without derivation (putting $c_\mathrm{I} = 1$)

$$\psi = c_\mathrm{I}\psi_\mathrm{I}+c_\mathrm{II}\psi_\mathrm{II} = \psi_\mathrm{I}+\frac{E-C-u+v}{u-w}\psi_\mathrm{II}. \quad (21)$$

There are two wave functions ψ corresponding to the two energy values E. c_{II}/c_I depends on the exchange integrals A and hence on the distances between the atoms.

The generalization of this method (Heitler, Rumer, Weyl, 1931) for atoms with more than one electron will be given in section 4.

2. Activation energy, non-localized bonds

We now consider the interaction of three hydrogen atoms in more detail. This will give us a new insight into the mechanism of the covalent bond and we shall see also that new concepts arise. We begin with some purely qualitative considerations which will be corroborated afterwards by the theory.

In order to distinguish the H atoms it is convenient to assume that one of the three atoms is in fact an atom of heavy hydrogen (or deuterium). Such a D atom behaves chemically exactly like H, its electron has the same wave functions and energy-levels as H, it differs from H only in that the mass of the nucleus is twice that of H. For simplicity we assume that the three atoms are always placed on a straight line and we only change the distances. Other cases can, of course, be treated equally well.

In the first place we assume that the two H atoms have formed a molecule and that the D atom approaches the H_2 molecule from a large distance: H—H···D. We have seen in Chapter IX, section 2, that then the D atom is repelled because there is no possibility of forming a pair with one of the H atoms. This is true as long as the distance H···D is much larger than the distance H—H of the molecule. But what happens when the D atom has come near to the H_2 molecule? Supposing we have brought the D atom to just about the same distance from the right-hand H atom as the equilibrium distance H—H.

POLYATOMIC MOLECULES 165

Will the D atom then still be repelled? We find the answer to this question if we now remove gradually the left-hand H atom: H···H—D. When the H atom is removed very far we are left with a system H—D. H—D can now either be in a triplet or in a singlet state, in which cases the atoms repel or attract each other respectively. At first, one might think that the former must be the case (repulsion), since when the D atom was far away it was repelled by the two H atoms. This, however, is not true, as can be seen as follows: We started from a state H—H···D which has an energy equal to the binding energy of H_2, and this is the lowest energy which the three atoms can have in their present arrangement. The spin of the whole system is $\frac{1}{2}$ and does not change. Now if we change the distances no interaction curve with the same total spin[†] can cross, and therefore, however we change the distances, the energy must remain the lowest possible for each given arrangement of the atoms (and spin $\frac{1}{2}$). If we bring the D atom near and afterwards remove the H atom,

$$H-H\cdots D \qquad H\cdots H-D,$$

we always remain in the lowest energy state, at each stage. Finally, we have an H—D system and its lowest state is the singlet (attraction). We are therefore left in the end with an HD molecule. In other words: *While changing the distances as above, the bond between* H—H *goes over into a bond between* H—D.

This changing over of a bond from one pair of atoms to another pair cannot, of course, happen suddenly. There must be intermediate states in which, so to speak, half a bond is established between HH and half a bond between HD:
$$--H--H--D--.$$

[†] The angular momentum is here always zero.

By symmetry reasons this will be especially the case when the HH and HD distances are equal. We are dealing here with a case where the *chemical bond* is *not localized* between pairs of atoms. The number of electron pairs formed between H and H is neither zero nor one (it is one when the D atom is far away, zero when H is removed); in fact, this number has in the intermediate state --H--H--D-- not a 'sharp value' (in our previous language). In Chapter IV, p. 53, we have seen that then the wave function is composed of parts, each with a *sharp number of pairs*, i.e. with *localized bonds*. In our case there are two parts, one with a pair localized between H—H and the other with a pair between H—D. This intermediate state is called a *transition state*.

How now does the energy change while we change the distances? From what was said above it is clear that at first the energy increases when the D atom approaches the H_2. Later, when the intermediate stage H--H--D is reached, the increase of energy will be less rapid as a bond between H and D is being built up (at the cost of the H—H bond, though). It may even be that in the stage H--H--D the energy has a small relative minimum. At any rate, the energy will decrease rapidly again if we remove the H atom, because we shall in the end be left with an HD molecule. Thus, if we change the distances as above, the energy follows a curve of the type of Fig. 40, where it is plotted qualitatively as a function of R_{HD}/R_{HH}. It has a *maximum*.

Our conclusions have an important bearing on the theory of chemical reactions. Supposing we study the exchange reaction $H_2+D = HD+H$. Although the energies before and after the reaction are exactly the same (namely, the binding energy of H_2) and the reaction therefore isothermal, we see from Fig. 40 that the reaction cannot take place unless a certain amount of energy, the so-called *activation energy*, is made available. For before the D atom can be attracted, it must be brought across the maximum of the curve of Fig. 40. The activation energy required is equal to the height of this maximum (in the figure called F). Naturally, F depends on the direction from which the D atom approaches, but qualitatively the curve is the same for all directions. F is lowest when the three atoms are on a straight line and then has the experimental value 0·35 eV.

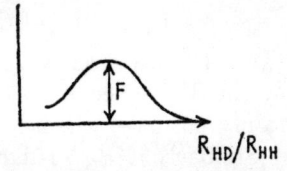

FIG. 40. Activation energy.

The same considerations as above can be applied to reactions between molecules, for instance, $H_2+D_2 = 2HD$, only the activation energy will be much larger because the repulsion between two saturated molecules is stronger than that between a molecule and a free atom.

This explanation of the activation energy and the more quantitative theory below was first given by London (1928).

These more intuitive considerations are fully borne out by the theory. The above conclusions are all contained in formula (20) for the interaction energy and (21) for the wave function. Since we have only three atoms, let the interaction integrals of one of the four atoms (c) vanish. The lowest energy is that where the square root occurs with negative sign. We identify the atoms a, b, d then with H, H, D, in the order of their arrangement. Call the integrals between the neighbours C_{HH}, A_{HH} and C_{HD}, A_{HD} respectively and those between the non-neighbours C', A'. C', A' are always rather small. Then

$$E = C_{HH}+C_{HD}+C'-$$
$$-\sqrt{\{A_{HH}^2+A_{HD}^2+A'^2-A_{HH}A_{HD}-(A_{HH}+A_{HD})A'\}}. \quad (22)$$

Consider first the case where R_{HD} is large and R_{HH} near the equilibrium distance. Then A_{HD} is also small and by expansion of the square root we find

$$E = C_{HH}+C_{HD}+C'-|A_{HH}-\tfrac{1}{2}(A_{HD}+A')|$$
$$= C_{HH}+A_{HH}+C_{HD}+C'-\tfrac{1}{2}(A_{HD}+A'). \qquad (23)$$

(The A's are negative.) $C_{HH}+A_{HH} \equiv E_{H_2}$ is the energy of the H_2 molecule. $C_{HD}-\tfrac{1}{2}A_{HD}$ and $C'-\tfrac{1}{2}A'$ are certainly positive, which means that the D atom is repelled.

Similarly if R_{HH} is large but R_{HD} near the equilibrium value,

$$E = C_{HD}+A_{HD}+C_{HH}+C'-\tfrac{1}{2}(A_{HH}+A'). \qquad (23')$$

An HD molecule is formed and the third H is repelled. The two cases go over into each other by a mere adiabatic change of the distances because they are both special cases of the same energy state represented by (22). When both distances are equal, $A_{HH} = A_{HD}$, for $R_{HH} = R_{HD} = R_0$, say, (22) reduces to

$$E = C_{HH}+A_{HH}+(C_{HH}+C'-A')$$
$$= E_{H_2}+(C_{HH}+C'-A'). \qquad (24)$$

The energy differs from that of a single H_2 molecule by $C_{HH}+C'-A'$. A glance at Fig. 34, p. 133, shows that $C'-A'$ (> 0) at the distance $2R_0$ is much larger than C_{HH} at R_0. Thus the energy is certainly higher than E_{H_2}. On the other hand, it is difficult to decide at exactly what distance R_{HD} the maximum of the energy lies. It may be that the energy increases steadily when $R_{HH} = R_0$ is kept fixed and R_{HD} is allowed to decrease down to the value R_0. Or it may be that the maximum lies at a slightly larger value of R_{HD}, in which case the position $R_{HD} = R_{HH} = R_0$ would represent a relative energy minimum. We have neglected too much (overlap integral, three-body forces, influence of higher states, etc.) to be able to predict such minor details.

We can represent E as a function of R_{HH} and R_{HD} (always in linear arrangement) by an energy surface. This is shown qualitatively in Fig. 41, where the lines of equal energy are plotted. The 'cheapest' way in which an exchange reaction $H_2 + D \to H + HD$ can be effected is along the dotted line. It leads from the valley on the

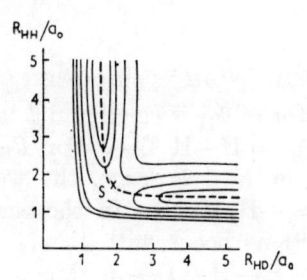

FIG. 41. Energy surface of the HHD system.

right ($R_{HH} = \infty$) across the saddle-point S, to the valley on the top ($R_{HH} = \infty$). The saddle-point may (or may not) have a little dip.† It is hard to calculate the activation energy F quantitatively, because it is the small difference of two large quantities (energy of three H atoms and binding energy of H_2) and depends sensitively on the values of the exchange integrals and the simplifications made. More than an agreement in order of magnitude cannot be expected, and this the theory achieves.

Finally consider the wave function. This is given by (21), section 1. ψ_I corresponds to the valency structure with a bond between the H atoms, i.e. to H—H D—. ψ_{II} is the structure —H H—D. In our present notation ψ is

$$\psi = \psi_I + \frac{E - C - A_{HH} + A_{HD}}{A_{HH} - A'} \psi_{II}.$$

† Actually S lies at a distance $R_{HH} = R_{HD}$ slightly larger than the equilibrium distance R_0 of H_2.

We obtain in the three cases (23)–(24),† (omitting A')

$$\psi = \psi_\mathrm{I} + \frac{1}{2}\frac{A_\mathrm{HD}}{A_\mathrm{HH}}\psi_\mathrm{II} \quad (R_\mathrm{HD} \text{ large}, A_\mathrm{HD} \text{ small}), \tag{25a}$$

$$\psi = \frac{1}{2}\frac{A_\mathrm{HH}}{A_\mathrm{HD}}\psi_\mathrm{I} + \psi_\mathrm{II} \quad (R_\mathrm{HH} \text{ large}, A_\mathrm{HH} \text{ small}), \tag{25b}$$

$$\psi = \psi_\mathrm{I} + \psi_\mathrm{II} \quad (R_\mathrm{HH} = R_\mathrm{HD}, A_\mathrm{HH} = A_\mathrm{HD}). \tag{25c}$$

In (25a) the factor of ψ_II is small and ψ is essentially the wave function $\psi_\mathrm{I} =$ H—H D—. For $R_\mathrm{HD} = \infty$ this is exactly so. When R_HH is large, the wave function is essentially $\psi_\mathrm{II} =$ —H H—D. In the case $R_\mathrm{HH} = R_\mathrm{HD}$ both wave functions occur with the same factor. The *transition state* --H--H--D-- is thus a *superposition of two valency structures*. The bonds are *not localized* and the number of pairs formed between a specified pair of atoms does not have a sharp value.

We shall see that the behaviour of three H atoms is characteristic for almost all polyatomic molecules.

3. Directed valencies

The valencies of many atoms with more than one valency have a bias to form *definite angles*. The tetrahedral structure of the four valencies of carbon is well known. Also the two valencies of oxygen extend in definite angles, the water molecule has the form of a triangle with a basic angle of 105°. NH_3 is a pyramid. These directional properties of the valencies of O, N, or C must be due to the structure of these atoms themselves. Supposing, for instance, that the O atom was spherically symmetrical, the two

---

† We have multiplied (25b) by $A_\mathrm{HH}/(2A_\mathrm{HD} - \frac{3}{2}A_\mathrm{HH}) \simeq A_\mathrm{HH}/2A_\mathrm{HD}$ in order to make the factor of ψ_II equal to unity. It is, of course, only the ratio of the two coefficients that matters.

POLYATOMIC MOLECULES

valencies would extend in any direction and the attraction forces between O and the two H atoms would be independent of the angle between them. The form of the H_2O molecule would then depend on the forces between the H atoms alone. Now O in its ground state has two electrons with parallel spin and the two H atoms must both have the same spin direction, namely opposite to that of O. The two H atoms therefore repel each other. It would follow that the most stable form of H_2O is a straight line: H—O—H. This is not found to be the case.

In our previous considerations we have chiefly concentrated our attention on the question whether or not a bond between two atoms can be established, and we have seen that this depends on the formation of pairs. We have not, so far, inquired to what extent the forces between two atoms, due to the formation of a pair, depend on the quantum states and wave functions of the two electrons concerned. Now there is an essential difference in the valencies due to electrons in s and p states. In particular we shall see that valencies of *p electrons have just the desired directional properties*.

As an example we consider the simplest case, the water molecule. We begin again with some qualitative considerations. The O atom, in its ground state (^3P) has, according to Chapter VII, section 2, four electrons in the $2p$ shell. For the wave functions of the three degenerate p levels we choose the functions given in (9) and (11) of Chapter III, namely (neglecting the interaction between the electrons),

$$\psi_x = xf(r), \qquad \psi_y = yf(r), \qquad \psi_z = zf(r). \qquad (26)$$

They are pictured in Fig. 12, p. 43. Of the four electrons two must be placed with antiparallel spin in one of these three states, say, for instance, in ψ_z. The other two electrons can then be placed in ψ_x and ψ_y respectively, and can have parallel spin. In the ground state ^3P they *have* parallel spin (see Fig. 42; only ψ_x and ψ_y are shown; ψ_z would extend in the direction perpendicular to the plane of drawing).

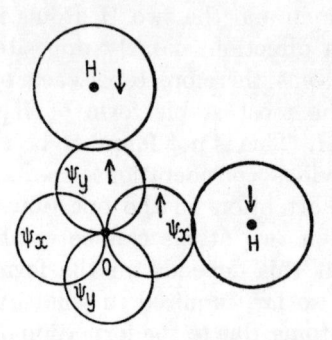

Fig. 42. Water molecule.

Now let a hydrogen atom be brought in contact with the oxygen atom. In Chapter IX it has been found that the interaction energy of two atoms (the integrals C and A) depends on the fact that the wave functions of the two atoms *overlap*. If they overlap very little the interaction energy is very small. We see at once that in our case the interaction must depend on the *direction* from which the H atom approaches the O atom. If the H atom comes from the z-direction, its wave function (which is spherically symmetrical) will overlap a great deal with that of the pair placed in ψ_z, but overlaps very little with the two electrons in ψ_x and ψ_y. The result must be that the H atom is repelled. Attraction arises if the H atom approaches either from the x or y direction. Then a pair can be formed between H and, say, the free electron in ψ_x. The effect of the pair in ψ_z will then be small. If the H atom comes from a direction between x and y, it is also attracted, but the attraction will be strongest from the x or y direction itself.

Now let the second H atom be brought near. If the first H atom has already formed a pair with the electron in ψ_x, then the direction from which the second H atom is most strongly attracted is the y-direction. The second H atom can form a pair with the electron in ψ_y. It follows that the most stable configuration of the water molecule is a triangle with basic angle of 90°:
$$\begin{array}{c} \text{H} \\ | \\ \text{O—H} \end{array}$$
. This will be slightly modified by the fact that the two H atoms repel each other. The angle will then be slightly larger than 90°, as is actually the case.

By permitting the electron in ψ_x to form a pair with the first H, and that in ψ_y with the second H atom, we actually permit these two electrons to arrange their spins independently. They cannot always have parallel spin (as in the ^3P state) because an exchange of spin direction with the hydrogen partner takes place. This means that, in addition to ^3P, we also take the singlet states of O (^1D, ^1S) into consideration. In fact we apply what in Chapter X, section 3 was called the procedure (b). It is in this approximation where the directional properties of the valencies are exhibited most directly.

In a very similar way the structure of NH_3 can be explained. The three p electrons of N can just be placed in the three states ψ_x, ψ_y, ψ_z. The three H atoms can all be attracted if they come from the x, y, z directions respectively. NH_3 should therefore have the form of a pyramid with an angle H—N—H of 90°. Again this pyramid will be flattened by the repulsion of the H atoms. The actual angle is 109°.

This explanation of the directional properties of valency was given by Pauling and Slater (1931).

These qualitative considerations can easily be expressed in formulae within the framework of our general theory.

Applying, thus, approximation (*b*) we consider the configuration of oxygen where two electrons are in ψ_z and one electron each in ψ_x and ψ_y.† We make no restriction as to their spin directions.

We consider then two H atoms placed either (i) both on the *x*-axis on both sides of the O atom (angle 180°) or (ii) one on the *x*-axis and one on the *y*-axis (angle 90°) and we shall show that the binding energy is greater in the latter case.

To calculate the interaction energy we can use the general formula for the interaction of four electrons derived in section 1. Although this formula was meant to apply to four atoms with one electron each it is valid for four electrons occupying any four different one-electron states. The nature of these states enters in the integrals only. All we have to do is to replace two of the 'atoms' by the two states ψ_x, ψ_y. Only one addition is required, namely the interaction of the two H atoms with the electron pair occupying ψ_z. It is very easy to see that this leads merely to an addition $C-A$ for each H atom where C, A are the interaction integrals of the H atom with an electron in ψ_z (compare (9) of Chapter IX, section 2).

We denote the interaction integrals between the first or second H atom with the electrons in ψ_x, etc., by C_{1x}, A_{1x},

† The electrons of oxygen can, of course, be arranged in various ways in the *p*-shell and all these configurations have in general to be taken into account (for example two electrons in ψ_x and two in ψ_z). In the two cases considered below where the H atoms are placed either at 180° or at 90° it can be shown from symmetry considerations that none of the other configurations enter the problem.

POLYATOMIC MOLECULES

$C_{2x},...,C_{2y}$, etc., those between the two H atoms by C_H, A_H. For simplicity we neglect the interaction between the electrons within the O atom (which is by no means necessary). Formula (20) with the abbreviations (18a) of section 1 becomes

$$E = C + C_{1z} + C_{2z} - A_{1z} - A_{2z} - \sqrt{\{u^2 + v^2 + w^2 - uv - uw - vw\}},$$

where

$$C = C_{1x} + C_{2x} + C_{1y} + C_{2y} + C_H,$$
$$u = A_{1x} + A_{2y}, \qquad v = A_{1y} + A_{2x}, \qquad w = A_H.$$

Consider now the two positions 180° and 90°, and let the distances of the H atoms from the oxygen be equal. For the 180° position it follows from symmetry reasons that

$$C_{1x} = C_{2x} \equiv C_0, \qquad A_{1x} = A_{2x} \equiv A_0,$$
$$C_{1y} = C_{2y} = C_{1z} = C_{2z} \equiv C',$$
$$A_{1y} = A_{2y} = A_{1z} = A_{2z} \equiv A'$$

and E becomes:

180°: $E = 2C_0 + 4C' + C_H - 2A' - |A_0 + A' - A_H|$
$\quad\quad = 2C_0 + 4C' + C_H + A_0 + A' - A_H \qquad (27a)$

($A_0 + A' - A_H$ is certainly negative because the distance between the H atoms is rather large and A_H small).

From the above overlapping considerations it follows that $|A_0|$ is rather large and $|A'|$ considerably smaller.† Consider next the 90° position. The second H atom is now on the y-axis. From symmetry reasons it follows that now

$$C_{2y} = C_{1x} = C_0, \qquad A_{2y} = A_{1x} = A_0,$$
$$C_{2x} = C_{1y} = C_{1z} = C_{2z} = C',$$
$$A_{2x} = A_{1y} = A_{1z} = A_{2z} = A',$$

† An explicit calculation shows that A' has, contrary to the general rule, a positive sign, whereas A_0 is, as usual, negative.

where A_0 and A' are the same integrals as in (27a). We obtain

$$90°: E = 2C_0 + 4C' + C_H - 2A' - \\ - \sqrt{\{4A_0^2 + 4A'^2 - 4A_0A' + A_H^2 - 2A_H(A_0 + A')\}}. \tag{27b}$$

Let us now compare (27a) with (27b). The 90° position has a lower energy if

$$4A_0^2 + 4A'^2 - 4A_0A' + A_H^2 - 2A_H(A_0 + A') \\ > (A_0 + A' - A_H)^2,$$

or if $\qquad (A_0 - A')^2 > 0.$

Now this is just the case on account of the larger overlapping in the integral A_0. *Thus the 90° position is more stable* than the 180° position. Since A_H is rather small, certainly compared with A_0, it is also seen from (27) that the two H atoms repel each other. (The term

$$-2A_H(A_0 + A')$$

in (27b) diminishes the binding energy and A_H^2 is smaller than this term.) The angle with the lowest energy is therefore slightly larger than 90°, as is in fact the case.

It is instructive to consider the same problem also when we take only the ^3P state into account (procedure (a)). The wave function of the two O electrons in ψ_x and ψ_y is then antisymmetric. If we were to neglect the interaction between the two electrons we should have

$$\psi(1,2) = -\psi(2,1) = \frac{1}{\sqrt{2}}\{\psi_x(1)\psi_y(2) - \psi_x(2)\psi_y(1)\}.$$

With the help of the explicit expressions (26) this can also be written

$$\psi(1,2) = \frac{1}{\sqrt{2}}(x_1y_2 - x_2y_1)f(r_1)f(r_2) = \frac{1}{\sqrt{2}}(\mathbf{r}_1 \times \mathbf{r}_2)_z f(r_1)f(r_2), \tag{28}$$

where $\mathbf{r}_1$, $\mathbf{r}_2$ are the position vectors of the two electrons. The exact wave function differs, of course, from (28) but the symmetry properties are the same. (28) has its largest value when the two vectors $\mathbf{r}_1$, $\mathbf{r}_2$ are perpendicular to each other. This fact, however, does not so far give rise to directional properties. The theory for the interaction of atoms with several electrons in antisymmetrical states will be developed in the next section. A very simple calculation shows that the energy is, apart from Coulomb terms and the contribution from the pair in ψ_z, just one exchange integral A for each H atom, where

$$A = \int \psi(1,2)\psi_{\mathrm{H}}(3) V \psi(1,3) \psi_{\mathrm{H}}(2) \, d\tau \qquad (29)$$

and ψ_{H} is the wave function of the H atom. The integral has the same value whether the H atom lies on the x or the y axis.[†] In fact it is independent of the direction from which the H atom approaches. The energy of H_2O is the sum of two such integrals (29) for each H atom. This is independent of the angle H—O—H. In fact, if one neglects the interaction of the O electrons, one obtains just the expression (27 a) we found before for the 180° position. The lowering of the energy in the 90° position is due to the fact that we have allowed the two O electrons to arrange their spins freely, which implies that singlet states of O are also taken into account. It is the influence of these (in reality excited) states which causes the directional preference and it is in accordance with our considerations in Chapter X, p. 147, that the energy is lowered.

In addition to the influence of the excited states there is a second effect which contributes to the directional properties. We have always confined ourselves to the exchange of only two electrons. It was mentioned in

[†] $\psi(1, 2)$ merely changes sign if x and y are interchanged and ψ occurs twice in (29).

section 1 that higher exchanges also occur. For example, if the two O electrons are numbered 1, 2, those of the H atoms 3, 4, we can have an exchange 1 ↔ 3 simultaneously with 2 ↔ 4. Such exchanges give rise to smaller contributions to the energy which, however, depend on the position of the three atoms simultaneously and not only on the distance between two atoms. If these are taken into account even the ^3P state alone gives preferential directions: The wave function (28) is largest when the two electrons are at right angles and the value of such a double exchange integral is also largest when the two H atoms form a right angle with O. (Artmann, 1946.)

Thus there are two factors contributing to the directional valencies of H_2O: (i) the influence of the excited states ^1D, ^1S, and (ii) double exchange, which is effective even in the ground state ^3P. For molecules like NH_3 the situation is very similar. The excited states arising from the configuration with three p-electrons give rise to a pyramidal structure but even the ^4S state, although it is spherically symmetrical as a whole, has a wave function which is largest when the three electrons are at right angles. Double exchange leads to the same structure.

The case of carbon is a little more complicated. It can be shown that the co-operation of ^3P and ^5S also leads to preferential directions, namely the directions from the centre of a tetrahedron towards its corners, although this cannot be shown by such intuitive considerations as for O or N. Furthermore, the wave function of ^5S is largest when the four position vectors r_1, r_2, r_3, r_4 extend in the directions of a tetrahedron. This is in fact quite easy to see. Thus the four valencies of carbon have a bias to form a tetrahedral structure.

Although the directional properties of the valencies are very important in that they determine the geometrical structure of a molecule, they are in many cases less important if we are interested in the binding energies. For example, the energy required to press the NH_3 pyramid flat is only 0·26 eV, a small fraction of the binding energy (in contrast to H_2O where the energy difference between the 90° and the 180° positions is large). Also for carbon the energy required to bend the valency angles is rather small compared with the binding energy. Both effects which give rise to the directional bias involve comparatively small energies. In the following sections, where the emphasis is on the binding energies, we shall ignore the directional properties (we can use what in Chapter X, section 3, was called the method (a)).

4. Interaction of atoms with several electrons

We now generalize the theory of section 1 to atoms with several electrons. Let a, b, c,... be atoms with n_a, n_b, n_c,... valency electrons. For simplicity we assume that no degeneracy other than that due to the spin exists. Further we assume that the n_a electrons of atom a have parallel spins so that atom a is in an $^{n_a+1}S$ state, and we do not take into account electron pairs within the atom. The orbital wave function of a is then antisymmetrical in the n_a electrons, the spin wave function is symmetrical. The simplifications are made solely for the purpose of laying the emphasis on what is most essential and interesting in the bond mechanism. It would be very easy indeed to take the degeneracy of P states and electron pairs, etc., into account. Thus we apply what in Chapter X, section 2 was called the procedure (a) and we have pointed out that this can even be applied to carbon (^5S state) for the neighbourhood of the energy minimum.

The modification of section 1 for the present case requires hardly any calculation. The first step is again the construction of the valency structures. This is much as in section 1, the only difference is that now more than one electron pair can be formed between atoms a and b. Suppose, for example, $n_a \geqslant n_b$, then evidently up to n_b pairs can be formed. Also, of course, an atom can form bonds with several atoms. For each such structure a spin wave function ϕ can be constructed. For example, if we have two C atoms and two H atoms the following three valency structures exist:

$$
\begin{array}{ccc}
C \Rrightarrow C & C \Rrightarrow C & C \Rrightarrow C \\
\downarrow \quad\; & \downarrow \quad \downarrow & \times \\
H \longrightarrow H & H \quad H & H \quad H \\
\phi_{\text{I}} & \phi_{\text{II}} & \phi_{\text{III}}
\end{array} \qquad (30)
$$

Again there is a linear relation between the structures relating to any four bonds involving four atoms. It is exactly the same as in section 1, for example,

$$\phi_{\text{III}} = \phi_{\text{II}} - \phi_{\text{I}}. \qquad (31)$$

There are no further linear relations. Again an 'independent basis' of spin wave functions is found by the non-crossing rule. The structures (30) are, of course, constructed without any prejudice as to the formation of a molecule C_2H_2.

The interaction integrals are now defined in the following manner: The Coulomb integral is, as before,

$$C = \int \psi_a^2 \psi_b^2 ... V \, d\tau = C_{ab} + C_{ac} + C_{bc} + ... \qquad (32)$$

The integration is to be extended over all valency electrons of all atoms and V is a sum of terms each of which describes the interaction between two atoms. The exchange integral, between atoms a and b, is

$$A_{ab} = \int \psi_a(1, 2, ..., n_a)\psi_b(n_a+1, ..., n_a+n_b) V \times$$
$$\psi_a(n_a+1, 2, ..., n_a)\psi_b(1, n_a+2, ..., n_a+n_b) \, d\tau. \quad (33)$$

We have numbered the electrons of a $1, 2, ..., n_a$, those of b $n_a+1, ..., n_a+n_b$, and (33) involves the exchange

$$1 \leftrightarrow n_a+1.$$

The integral has the same value if any other two electrons belonging to a and b are exchanged. There are also exchange integrals in which higher permutations involving more than two electrons occur. These can now even occur for two atoms (exchange, for example, two electrons of a against two electrons of b), and there are, of course, also exchanges in which three or more atoms take part. All these are considerably smaller than the ordinary exchange integral and we shall neglect them.

The interaction energy is again determined from the set of equations (for N independent valency structures)

$$(E-C)\phi_\mathrm{I} + \sum_{a,b} A_{ab} T_{ab} \phi_\mathrm{I} = 0$$
$$\cdots \cdots \cdots \cdots \quad (34)$$
$$(E-C)\phi_N + \sum_{a,b} A_{ab} T_{ab} \phi_N = 0.$$

The next question is: what is the effect of the exchange T_{ab} on ϕ_I, etc.? Evidently, T_{ab} is the sum of interchanges of the end-points a and b of any two valency dashes ending in (or starting from) a and b. If p_{ab} bonds connect a with b the corresponding contribution is $T_{ab} \phi = -p_{ab} \phi$. If a is connected with c by p_{ac} bonds, b with d by p_{bd} bonds, then $p_{ac} p_{bd}$ exchanges can take place and the result is $p_{ac} p_{bd} \phi'$, where ϕ' is the structure which has $p_{ac}-1$ bonds a—c, $p_{bd}-1$ bonds b—d, and one bond each b—c and a—d. It is better to express this in formulae. We symbolize a bond a—b by $[ab] = -[ba]$. Then the structure with p_{ab} bonds a—b, p_{ac} bonds a—c, etc., is denoted by

$$\phi = [ab]^{p_{ab}}[ac]^{p_{ac}}[bd]^{p_{bd}}\ldots.$$

The linear relation of type (31) is expressed by

$$[ad][bc] = [ac][bd]-[ab][cd].$$

Then $T_{ab}\phi$ is given by

$$T_{ab}\phi = -p_{ab}\phi + \sum_{c,d} p_{ac}p_{bd}\phi \frac{[ad][bc]}{[ac][bd]}. \quad (35)$$

The sum is to be extended over all atoms c, d other than a, b. It may also be that c and d are the same atom (if both a and b are connected with the same third atom c). The structure in the second term of (35) has one additional bond a—d and b—c in place of a bond a—c and b—d.

We give two examples. Consider the structures ϕ_I, ϕ_II (30) which will describe the C_2H_2 molecule. For the exchange T_CC between two C atoms we have

$$T_\text{CC}\phi_\text{I} = -4\phi_\text{I},$$
$$T_\text{CC}\phi_\text{II} = -3\phi_\text{II}+\phi_\text{III} = -2\phi_\text{II}-\phi_\text{I}. \quad (36\,a)$$

If C, H are two atoms connected by a bond in ϕ_II then

$$T_\text{CH}\phi_\text{I} = -4\phi_\text{III} = -4\phi_\text{II}+4\phi_\text{I}, \; T_\text{CH}\phi_\text{II} = -\phi_\text{II}. \quad (36\,b)$$

We can now calculate the energy of the C_2H_2 system. Let the atoms connected by a bond in ϕ_II be neighbours (ϕ_II is then the chemical formula of the C_2H_2 molecule), and let us neglect the interaction of non-neighbours. The C—H interaction occurs twice, so C_CH and A_CH have to be multiplied by a factor 2. Inserting (36) into (34) we obtain

$$(E-C)\phi_\text{I}+(-4A_\text{CC}+8A_\text{CH})\phi_\text{I}-8A_\text{CH}\phi_\text{II} = 0,$$
$$-A_\text{CC}\phi_\text{I}+(E-C)\phi_\text{II}+(-2A_\text{CC}-2A_\text{CH})\phi_\text{II} = 0.$$

It follows that

$$\begin{vmatrix} E-C-4A_{\text{CC}}+8A_{\text{CH}} & -8A_{\text{CH}} \\ -A_{\text{CC}} & E-C-2A_{\text{CC}}-2A_{\text{CH}} \end{vmatrix} = 0.$$

The solution of this quadratic equation for E can be written in the form (with $C \equiv C_{\text{CC}}+2C_{\text{CH}}$)

$$E = C_{\text{CC}}+3A_{\text{CC}}+2(C_{\text{CH}}+A_{\text{CH}})-A_{\text{CC}}f_3\left(\frac{A_{\text{CC}}}{2A_{\text{CH}}}\right),$$

$$f_3(\xi) = \frac{5}{2\xi} \pm \frac{1}{\xi}\sqrt{\left[\xi(\xi-1)+\frac{25}{4}\right]}, \qquad \xi \equiv \frac{A_{\text{CC}}}{2A_{\text{CH}}}. \quad (37)$$

We shall use and discuss this formula in detail in section 5.

As a second, very simple, example consider two atoms with n and m valency electrons respectively. Assume $n \geqslant m$ and consider also the case where only p pairs are formed, $p \leqslant m$, so that atom a has $n-p$, atom b $m-p$ free valencies. To reduce this case to the above we introduce a third atom, L say, which has no interaction with a and b (i.e. $C_{aL} = A_{aL} = C_{bL} = A_{bL} = 0$) but which can absorb the free valencies of a and b. There is then only one valency structure, namely

$$\phi = \underset{n-p\;L\;m-p}{\overset{a \overset{p}{=\!=\!=\!=} b}{\diagup\!\!\!\diagdown}} = [ab]^p[aL]^{n-p}[bL]^{m-p}. \quad (38)$$

The application of (35) gives at once

$$T_{ab}\phi = -p\phi+(n-p)(m-p)\phi.$$

The exchange of the end-points a, b of two valency dashes connecting a with L and b with L leads back to the same structure. Thus the energy is

$$E = C+[p-(n-p)(m-p)]A.$$

This is the formula anticipated and discussed in Chapter X, section 3.

5. Binding energies of hydrocarbons

The theory of the preceding section allows one to calculate the binding energy of a polyatomic molecule in terms of the Coulomb and exchange integrals. There is, of course, little hope that we could calculate these integrals theoretically without excessive labour, as could be done for hydrogen. Instead, a semi-empirical method suggests itself. If we have a number of different molecules consisting of the same kind of atoms so that in each molecule the same interaction integrals occur we can use the experimental data of some of the molecules to determine the integrals and then check the theory by calculating the energies of the remaining molecules and compare them with the experimental data. A particularly favourable case is available in the simple organic molecules consisting of C and H atoms only. Thus we shall consider the molecules CH_4, C_2H_2, C_2H_4, C_2H_6 and the radicals obtained by removing an H atom CH_3, C_2H, C_2H_5.

It has been explained in Chapter X, section 2, that for the treatment of molecules containing carbon it is a feasible approximation to consider the ^5S state of carbon with its four valencies alone, provided that we confine ourselves to the energy minimum and leave out of consideration molecules in which only one or two valencies operate (CH, CH_2). The binding energy thus obtained is then, however, the energy required to separate the atoms such that each C atom is left in the ^5S state. In reality the molecule dissociates into atoms in their ground state (^3P for carbon) and this dissociation energy is also what is measured. To obtain this quantity we have to subtract from the positive dissociation energy (= minus the binding energy) obtained from our formulae the excitation energy of ^5S, viz. $\nu\epsilon$, where ν is the number of C atoms in the molecule, and ϵ the excitation energy of ^5S.

The interaction integrals which occur are all negative and so is the binding energy. It is more convenient to deal with positive quantities. So let $D = -E$ now be the positive dissociation energy into atoms in their ground states. In the molecules CH_4, C_2H_6, and C_2H_4 the interaction between neighbouring H atoms (in the same CH_3 and CH_2 group) must be taken into account. To simplify writing we introduce the following notation:

$$-C_{CC} = c, \qquad -A_{CC} = \gamma,$$
$$-C_{CH} = b, \qquad -A_{CH} = \beta,$$
$$+(C_{HH} - A_{HH}) = h.$$

h is the interaction energy of two H atoms in the repulsive state (calculated in Chapter IX, section 1). $h > 0$.

The energies of CH_4 and CH_3 are easily calculated. There is only one valency structure identical with the chemical formula. The spins of the H atoms are all parallel. Thus the H atoms repel each other. The repulsion occurs six times in CH_4 and three times in CH_3. We obtain

$$CH_4: \quad D = -\epsilon + 4(b+\beta) - 6h, \qquad (39a)$$

$$CH_3: \quad D = -\epsilon + 3(b+\beta) - 3h. \qquad (39b)$$

The energy of C_2H_2 has already been calculated in the preceding section. We have to subtract the excitation energy of ^5S twice because there are two C atoms.

$$C_2H_2: \quad D = -2\epsilon + 2(b+\beta) + c + 3\gamma - \gamma f_3(\xi), \qquad (40)$$

$$f_3(\xi) = \frac{5}{2\xi} - \frac{1}{\xi}\sqrt{\left[\xi(\xi-1) + \frac{25}{4}\right]}, \qquad \xi = \frac{\gamma}{2\beta} \qquad (40')$$

(the minus sign of the square root gives the lowest state of the molecule). The energy of C_2H is obtained by replacing $2b$ by b and 2β by β because there is only one C—H interaction:

$$C_2H: \quad D = -2\epsilon + b + \beta + c + 3\gamma - \gamma f_3(2\xi). \quad (41)$$

The molecules C_2H_4 and C_2H_6 are more complicated. One can safely assume that the H atoms in each CH_2 or CH_3 group have parallel spin and can be treated like an atom with two or three valencies respectively (it can be shown that this incurs only a negligible error). The reader will easily convince himself that then three structures exist for C_2H_4 and four for C_2H_6. The resulting equations of third and fourth order can readily be solved numerically, because the energy depends, apart from certain additive constants, on the ratio $\xi = \gamma/2\beta$ only. The result is:

$$C_2H_4: \quad D = -2\epsilon + 4(b+\beta) - 2h + c + 2\gamma - \gamma f_2(\xi), \quad (42)$$

$$C_2H_6: \quad D = -2\epsilon + 6(b+\beta) - 6h + c + \gamma - \gamma f_1(\xi). \quad (43)$$

The functions f_1, f_2, f_3 are plotted in Fig. 43 for the range of values of $\gamma/2\beta$ which come into question. The energy of the radical C_2H_5 is obtained by replacing $6b$ and 6β by $5b$ and 5β and $6h$ by $4h$. (There are $3+1$ H—H neighbours.)

$$C_2H_5: \quad D = -2\epsilon + 5(b+\beta) - 4h + c + \gamma - \gamma f_1(\tfrac{6}{5}\xi). \quad (44)$$

If it were not for the functions f the formulae would have a natural and simple significance. Each bond of the chemical formula contributes just one corresponding exchange integral and, of course, the Coulomb integral. In addition two neighbouring H atoms contribute a repulsion $-h$. But the functions f are by no means zero and the result is more complicated.

One observation about C_2H_6 may be of interest. If we

separate the two CH_3 groups (c, γ small) f_1 tends to the value $f_1 = 3$. The energy is then

$$D = 2D_{CH_3} + c - 2\gamma,$$

where D_{CH_3} is the energy of the CH_3 radical. The exchange integral between the C atoms occurs with a factor -2. Although each CH_3 group has a free valency the two radicals repel each other at large distances. Thus a (probably very small) activation energy has to be overcome to form the C_2H_6 molecule from the radicals. On the other hand two CH_2 or CH groups attract each other strongly.

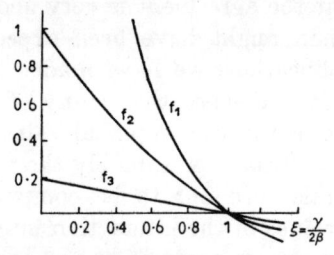

FIG. 43. The functions f_1, f_2, f_3 occurring in the energy of C_2H_2, C_2H_4, C_2H_6.

To use our results numerically we first remark that $\epsilon = 4\cdot 16$ eV is known. The interaction between H atoms is known both from theory and empirically. For the known H—H distance $h = 0\cdot 5$ eV. We are left with the four unknown constants b, c, β, γ.

We try to adjust these to fit four of the experimental data. It turns out, however, that the theoretical results depend in practice on three parameters only. So only three data can be fitted, for which we choose CH_4, C_2H_2, and C_2H_6. The remaining four dissociation energies are then determined practically whilst for the constants themselves a certain range of values is still possible. The test of the theory is therefore rather severe. In Table 11 the theoretical dissociation energies together with their experimental values† are given for the following choice of the constants (in eV) $c = 1\cdot9$, $\gamma = 3\cdot8$, $b = 3\cdot35$, $\beta = 2\cdot75$.

TABLE 11. *Dissociation energies of organic molecules*

Energy D	CH_4	C_2H_2	C_2H_4	C_2H_6	$CH_4 \rightarrow CH_3$	$C_2H_2 \rightarrow C_2H$	$C_2H_6 \rightarrow C_2H_5$
Exp.	17·2	17·0	23·5	29·1	4·45	5·25	4·25
Theor.	(17·2)	(17·0)	23·7	(29·1)	4·55	5·55	4·20

It is seen that the agreement is very good indeed, and much better than might have been expected from the numerous simplifications we have made. This can only be understood in the sense that many of the neglected effects are embodied in the empirical values of the interaction constants. This is presumably the reason why the Coulomb integrals turn out to be comparatively large. Otherwise the values of the interaction integrals are very reasonable.

The result shows that the *theory is well able to account for the so-called bond-energies of the single, double, and triple bond*, which occur in the energies of C_2H_2, C_2H_4, and C_2H_6. It must be realized, however, that the binding energies of these molecules are not additively composed of contributions due to each bond (with the exception of CH_4 and CH_3). Even if we take into account the excitation of the ^5S state ϵ

† The experimental values depend on the heat of evaporation of diamond which is still not finally known. The figures in the table refer to a value of 7·4 eV (= 170 kcal./mol.). An alternative value which cannot be entirely excluded is 6·1 eV. If this should be the true value, the experimental values of the first four columns would be smaller by 1·3 eV per C atom. In this case a similar adjustment of the constants can be made and the agreement is about equally good but no agreement could be reached for a substantially different value.

POLYATOMIC MOLECULES

and the repulsion of the H atoms h, no additivity prevails because of the presence of the functions f_1, f_2, f_3 which depend on the ratio $\gamma/2\beta$ and are far from zero. The theory also accounts well for the differences of the energy required to remove an H atom in the various molecules. The fact that these are not equal again shows that the binding energies are not additive.

The non-additivity of the binding energies is closely connected with the valency structures. As in the case of three H atoms (section 2) the wave function is (with the exception of CH_4 and CH_3) a superposition of several structures. In the simplest case, C_2H_2, these are the structures ϕ_I, ϕ_{II}, (30). The bonds are therefore *not localized* between pairs of atoms (compare section 2). For the above values of the energy and the constants one finds that

$$\phi = \phi_{II} + 0.165\phi_I.$$

The coefficient of ϕ_I is rather small. This is a very satisfactory feature: ϕ_{II} is just the structure of the *chemical formula* H—C≡C—H. Quite generally the structure of the chemical formula is always predominant amongst the various structures which occur in the wave function. This fact provides the *physical justification for the formulae* which have always been *used in chemistry* (especially in organic chemistry) with such outstanding success. Nevertheless, the other structures also play an important part and contribute very considerably to the energy. They are also necessary to obtain the correct asymptotic behaviour for large distances (compare section 2).

There are molecules where the bonds are not even nearly localized. The benzene ring with the two equivalent formulae ⌬ and ⌬ has always been a problem of organic chemistry. In wave mechanics it is obvious that, for symmetry reasons, both structures must occur in equal proportions in the wave function (as was first noticed by Pauling). Thus the double bond has no localized position in the ring.

Reviewing the contents of the last three chapters it can be said that wave mechanics is the tool for a complete understanding, on a physical basis, of all the fundamental facts of chemistry.

INDEX

Activation energy, 167 ff., 187.
Additivity of binding energies, 188.
Angular momentum, 50 ff.
— orbital, 61.
— spin, 59.
— total, 62.
— of molecule, 109, 113.
Antisymmetrical wave function, 75, 81, 83, 118, 126, 157.
Atomic interaction curves, 108, 143.

Balmer formula, 38.
Benzene, 189.
Binding energy, *see* Dissociation energy.
Bohr radius, 35, 131.
Boundary conditions, 22, 34.

C atom, excited states, 105.
—, ^5S-state, 104, 142.
—, valency, 142, 147.
C_2 molecule, 148.
Charge density, 89, 98, 125.
Closed shell, 99, 137, 142.
Complex wave functions, 18, 29, 44, 50.
Constants of the motion, 27, 51, 54, 109.
Continuous spectrum, 39.
Coulomb field, 24.
— energy, interaction, integral, 89, 130, 152, 160, 180, 185.
Coupling, spin-orbit, 63, 66.
—, two spins, 66, 92.
Crossing of atomic interaction curves, 143 ff.
CH_4, CH_3, C_2H_2, etc., *see* Hydrocarbons.

Degeneracy, 40, 44, 74, 95, 144.
— accidental, 147; *see also* Exchange degeneracy.
Deuterium, 164.
Diffraction, 2.
Directional properties of valencies, 171 ff.
— quantization, 57, 60.
Discrete quantum states, 1, 21.
Dispersion, 6.
Dissociation energy, of H_2, 133.
— of hydrocarbons, 188.
Double bond (C_2H_4), 188.
Doublets, 66, 110.
d, D-states, 42, 57, 67.
Δ-states, 110.

Effective charge of the nucleus, 98.
Eigenfunctions, 23.
Eigenvalues, 23.
— of hydrogen, 34, 37, 42.
Electron configuration, distribution, 73, 77, 100 ff.
— pairs (chemical bond), 133, 140, 149.
Equilibrium distance, of H_2 molecule, 133.
Exchange charge density, 90, 130.
— degeneracy, 75, 126.
— energy, integral, 89, 128, 153, 161, 181, 185.
Excited states, of H atom, 37 ff.
— —, of C, 105.
— influence on valency, 149, 151.
— on directional properties, 177.
Exclusion principle, 76 ff., 82, 117.

Fine structure, 64.
Forbidden states, 78.
Free electron, 17, 29, 37.

INDEX

Frequency, 1, 4, 7.
— of vibration of H_2, 134.
—— in diatomic molecules, 116.
— of spin exchange, 135.

Ground states, 34, 104, 141.
Group velocity, 5, 30.

Hamiltonian, 26.
Helium atom, 72, 88, 93.
—, interaction with hydrogen, 137.
Heterpolar bond, 123.
Homopolar bond, 123 ff.
Hydrocarbons, 184 ff.
— dissociation energy, 188.
Hydrogen atom, 32 ff.
— molecule, 124 ff.
——, ortho, para, 120.

Interaction energy, of two H-atoms, 133.
—, of two electrons, 91 ff.
— in diatomic molecules, 152, 183.
— of several H-atoms, 163.
— in H_2O, 175.
— in polyatomic molecules, 180 ff., 186.
Interference, 2.

Ionic bond, 123.
Ionization energy, 34, 106.

Lewis's electron pairs, 142.
Light quanta, 2.
Linear combinations of wave functions, 43, 95, 146.
Lithium, electron states, 78.
Localized bonds, *see* Non-localized bonds.

Magnetic dipole, 55.
— moment, 55, 60.
Measurement, 15.
Modulated wave, 30.

Molecule, electronic states, 107 ff., 154.
— rotation, 112 ff., 119.
— vibration, 115 ff.
Momentum, 4, 12.
Monochromatic waves, 12, 17.

Neutron, 120.
Non-localized bonds, 166, 170, 189.
Normalization, 43, 90, 96.
Nuclei, 120.
N_2, 121, 140, 154.
NH_3, 173.

Operator, 26, 47.
Ortho-molecules, 120.
Orthogonality relation, 93, 96.
Oscillator, 116.
Overlap integral, 129, 160.
O_2 molecule, 121, 148.

Para-molecules, 120.
Particle velocity, 4, 6, 31.
Pauli-principle, *see* Exclusion principle.
Periodic system, 99 ff., 141.
Perturbation energy, 88, 159.
Phase velocity, 4.
Potential barrier, 36.
Probability, 9 ff., 16, 35, 70.
p, P-states, 39, 43, 52, 67, 171.
Π-states, 110.

Quantum numbers, 38, 61.
— of action, 1.

Race, 111, 146.
Rare gases, 100, 137.
Repulsive interaction, 109, 132, 185.
Rotational levels (molecule), 114, 120.

Saturation, 123, 138.
Schrödinger equation, *see* Wave equation.
Screening, 98.

INDEX

Sharp and unsharp quantities, 10, 13, 47, 54, 166.
Singlets, 66, 82.
Size of an ion, 138.
Spin, 59 ff.
—, magnetic moment, 60.
—, of molecule, 110.
— valency, 139 ff., 149.
— wave function, 80, 96.
Standing waves, 17, 48.
Superposition of waves, 29, 49.
— of valency structures, 170, 189.
Symmetrical wave functions, 75, 81, 118, 126.
s, S-states, 39, 52, 67.
Σ-states, 110.

Three-body forces, 161.
Time-dependent wave equation, 28 ff., 49.
Transition state, 166.
Triple bond (C_2H_2), 188.
Triplets, 66, 82, 110.

Uncertainty relation, 13, 36, 117.

Valency of an atom, 141, 149.
— structure, 157, 169, 180.
Van der Waals' forces, 123.
Vector addition rule, 63, 67.
Vibrational levels (molecule), 116.

Water molecule, 171 ff.
Wave amplitude, 8.
— equation, 18 ff., 24 ff., 28.
— function of hydrogen, 33, 38, 40.
— — of oscillator, 116.
— — of s, p, d states, 40, 42, 52, 119, 171.
— length, 3, 7, 29.
— packet, 5, 49.

Zeeman effect, 55, 60.
Zero-point energy (vibration), 117.

PRINTED IN
GREAT BRITAIN
AT THE
UNIVERSITY PRESS
OXFORD
BY
CHARLES BATEY
PRINTER
TO THE
UNIVERSITY